LOGICAL PHYSICS

BOSTON STUDIES IN THE PHILOSOPHY OF SCIENCE

EDITED BY ROBERT S. COHEN AND MARX W. WARTOFSKY

VOLUME 74

A. A. ZINOV'EV

LOGICAL PHYSICS

Translated from the Russian by O. A. Germogenova

Edited by

ROBERT S. COHEN

D. REIDEL PUBLISHING COMPANY

A MEMBER OF THE KLUWER ACADEMIC PUBLISHERS GROUP

DORDRECHT / BOSTON / LANCASTER

Library of Congress Cataloging in Publication Data

Zinoviev, Aleksandr, 1922–
Logical physics.

(Boston studies in the philosophy of science; v. 74)
Translation of: Logicheskaia fizika.
Includes index.
1. Logic. 2. Physics–Philosophy. I. Cohen, Robert Sonné.
II. Title. III. Series.
Q174.B67 vol. 74 [BC57] 001'01s [530'.01] 83-15970
ISBN 90-277-0734-0

Published by D. Reidel Publishing Company,
P.O. Box 17, 3300 AA Dordrecht, Holland.

Sold and distributed in the U.S.A. and Canada
by Kluwer Academic Publishers
190 Old Derby Street, Hingham, MA 02043, U.S.A.

In all other countries, sold and distributed
by Kluwer Academic Publishers Group,
P.O. Box 322, 3300 AH Dordrecht, Holland.

Originally published in Russian by Nauka under the title: *Logičeskaja fizika*

Printed in The Netherlands

TABLE OF CONTENTS

EDITORIAL PREFACE xi

PREFACE xv

CHAPTER ONE / THE GENERAL THEORY OF INFERENCE AND TERMS

1. The Rules of Logic 1
2. Terms 2
3. Two Types of Terms 4
4. Simple and Complex Terms 4
5. Complex Terms 5
6. Occurrence of Terms and Statements in Other Terms and Statements 6
7. Metaterms and Metastatements 6
8. The Meaning of Terms 8
9. Terms Including Statements 9
10. Definitions 9
11. Statements 11
12. The Meaning of Statements 13
13. Definitions with Statements 13
14. The Definition of Predicates 15
15. The Truth Values of Statements 16
16. The Number of Truth Values 17
17. The Coordinates of Statements 18
18. Truth Value for Statements with Conjunction, Disjunction and Negation Operators 20
19. Truth Values of Other Forms of Statements 21
20. Tautology, Contradictions, Realizable Statements 24
21. Deduction 25
22. Logical Inference 26
23. The General Theory of Deduction 28
24. Classical and Non-Classical Cases in the Theory of Inference 31
25. The Rules of Inference and the Truth Values of Statements 32

26. Identity by Meaning and Entailment 33
27. The General Theory of Terms 33
28. The Coordinates of Statements 41
29. Consequences of Definitions 44
30. Implicit Definitions 45
31. Incomplete Definitions 46
32. Pseudodefinitions 47
33. Operational Definitions 48
34. Intuitively Obvious Statements 49
35. Variables 49
36. Definitions with Variables 53
37. The Multiple Meanings of Linguistic Expressions 54
38. Explication 55
39. Consistency of Terms 56

CHAPTER TWO / THE SPECIAL THEORY OF INFERENCE AND TERMS

1. The Logical Explication of Terms 57
2. Empirical and Abstract Objects 57
3. Individuals 59
4. Classes (Sets) 60
5. Clusters 63
6. States, Events 65
7. Existence 65
8. Quantifiers and Existence 68
9. Modal Predicates 69
10. Possibility 70
11. Contingent Events 73
12. Fatalism 75
13. Modal Operators 77
14. The Actual, the Existential, the Potential 77
15. The Measurement of Possibility 78
16. Relations 80
17. Comparison Relations 80
18. Order Relations 83
19. The Relation 'Between' 85
20. The Existence of Relations 85
21. Ordered Series 86
22. Contact 88

23. Continuity and Discontinuity of an Empirical Series 89
24. The Beginning and End of a Series 89
25. Interval 92
26. Length 92
27. Abstract Series 95
28. Finite and Infinite Series 96
29. Structure 97
30. Existence of a Structure 101
31. The Dimensions and Order of Structures 102
32. Correspondence 102
33. Correspondence of Classes 104
34. Functions 105
35. Ordered States 106
36. Conditional Statements with an Order-Relation 107
37. Functional Dependence 107
38. Connections 108
39. Ordering of Classes 110

CHAPTER THREE / NUMBER, MAGNITUDE, QUANTITY

1. Numbers in the Language 112
2. Numbers as Terms 113
3. Base Arithmetic 115
4. The Abbreviated Form of NB 117
5. The Universal Nature of Arithmetic 117
6. Expanded Arithmetic (EA) 118
7. Infinite Numbers 119
8. Formal Arithmetic with Metastatements 120
9. Formal Arithmetic and the Theory of Numbers 121
10. Number-Terms 123
11. The Existence of Numbers 123
12. Number as a Part of a Subject 125
13. Quantity 125
14. The Degrees and Range of Truth 127
15. Measurement and Definition 128
16. Number-Quantifiers 128
17. Amount 129
18. The Standard Classes of Numbers 129
19. The Cardinality of Classes of Numbers 130

20. Comparison of Cardinalities of Classes 131
21. Other Definitions 132
22. Reducibility to Logic 133
23. A Remark on the Class of Positive Integers 134
24. A Remark on a Paradox with the Number Terms 135

CHAPTER FOUR / LOGICAL PHYSICS

1. On the Method of Presentation of Logic 136
2. Empirical Individuals 137
3. Variation 138
4. The Transition State 140
5. Space and Time 140
6. Spatial and Temporal Relations 142
7. The Time and Space of Existence of an Empirical Individual 146
8. The Existence of Space and Time 149
9. The Spatial and Temporal Location of an Individual 151
10. One and the Same Individual 151
11. Change in Space and Time 155
12. The Irreversibility of Time 159
13. On the Relation of Generation 161
14. The Continuity of Space and Time 162
15. Invariance of Space and Time 164
16. Identity and Difference of Location and Time 164
17. The Predication of Variations 165
18. Translation, Change of Place (Movement) 167
19. The Paradox of Motion 168
20. Process 170
21. Minimal Dimensions 171
22. On Infinite Dimensions 176
23. Rate 181
24. Zeno's Paradoxes 185
25. Quanta of Space, Time and Motion 186
26. The Relativity of Motion 186
27. On the Existence and Motion of Clusters 191
28. The Beam 193
29. The Universe as a Whole 194
30. Empirical Geometry 197

31. Empirical Connections 204
32. The Predicates of Tendencies 205
33. Paradoxes of Connections 207
34. Conditional Predicates 209
35. Influence 210
36. Cause 214
37. Forms of Causal Relations 219
38. Determinism and Indeterminism 222
39. Other Forms of Connections 224
40. On the Logical Situation in Microphysics 225
41. The Wave-Particle Duality 232
42. The Trajectory 235
43. The Part and the Whole 235
44. Forecasts 237
45. The Hypothetical Ontology 239
46. General Assertions Concerning the Universe and Physical Assumptions 247
47. Generalizations of the Results of Science 249

APPENDIX

I. Epistemic Expressions 251
II. On the Concept of Belief 259
III. On the Notion of Preference 259
IV. On the Logic of Norms and Questions 262
V. On the Logic of Estimates 262
VI. On the Theory of Proof 265
VII. On Complete (Rigorous) Induction 266
VIII. On Logical Consistency 270
IX. On the Systems of the Logic of Classes 274

GLOSSARY OF SYMBOLS 279

SUBJECT INDEX 280

EDITORIAL PREFACE

In this stimulating study of the logical character of selected fundamental topics of physics, Zinov'ev has written the first, and major, stage of a general semantics of science. In that sense he has shown, by rigorous examples, that in certain basic and surprising respects we may envision a reducibility of science to logic; and further that we may detect and eliminate frequent confusion of abstract and empirical objects. In place of a near chaos of unplanned theoretical languages, we may look toward a unified and epistemologically clarified general scientific language. In the course of this work, Zinov'ev treats issues of continuing urgency: the non-trivial import of Zeno's paradoxes; the residually significant meaning of 'cause' in scientific explanation; the need for lucidity in the conceptions of 'wave' and 'particle', and his own account of these; the logic of fields and of field propagation; Kant's antimonies today; and, in a startling *aperçu*, an insightful note on 'measuring' consciousness.

Logical physics, an odd-appearing field of investigation, is a part of logic; and as logic, logical physics deals with the linguistic expressions of time, space, particle, wave, field, causality, etc. How far this may be taken without explicit use of, or reference to, empirical statements is still to be clarified, but Zinov'ev takes a sympathetic reader well beyond a realist's expectation, beyond the classical conventionalist.

Zinov'ev presents his investigations in four chapters and an appendix of technical elucidation. Chapter One, on the general theory of inference and terms, serves to express and clarify the author's conception of the principles and rules of logic which are requisite for proper analysis and deductive structuring of all expressions and all terms taken quite generally. With a rapid sweep from his early discussion of terms and statements, of the standard and non-standard notions of inference, and of elementary operators, Zinov'ev brings us to his analysis of definitions: implicit, incomplete, pseudo-, operational, and standard, and then to his discussion of multiple meanings. With Chapter Two, his special theory of inference and terms, Zinov'ev attacks the logic of terms (he sees all terms as designating objects), and we find that anything may be an object: empirical and abstract. Now we have a dense but lucid exposition of sets, and of

Zinov'ev's 'clusters', and then of the modal operators. The cluster is especially interesting with the property such that we may speak about spatial relations – dimensions, motions, displacements etc., – with reference to clusters (but not ordinarily with reference to classes or sets); and thereby we may expect Zinov'ev to set forth a distinctive notion of 'order' and 'succession' and such. Chapter Three deals with the central notions of arithmetic. With Chapter Four, the title essay on logical physics itself, Zinov'ev applies his logical apparatus to empirical individuals in their spatial and temporal relations. Among the interesting and perhaps striking sections are those on the 'minimal length', on irreversibility of time, on the fundamental 'beam', on the universe as a whole, on causal relations, on the logic of microphysics, on forecasts, and finally on the status of statements about the universe as a whole and in its fundamental nature. Such considerations of 'hypothetical ontology' as Zinov'ev provides (pp. 239–248) will perhaps be of greatest interest to philosophers as we reflect on this treatise in its many arguments and elucidations. And he has made it clear that the analytic explication of the logical structure of general hypotheses about the Universe (however much such logical analysis may be demanded if we are to be reasonable) must "not be confused with their acceptance". And yet logic helps to sort out the cases: statements for which it is impossible to check the extent of their truth; statements for which logic is competent to accept or reject them; statements for which logic is competent to say they are uncertain. How surprising and important the results will be to scientists and philosophers may be estimated from at any rate one result: "assertions about minimal dimensions and durations and maximal velocities represent logical assertions, while those about the finiteness and the infinity of the Universe are extralogical ones, although it seems [intuitively] that the opposite should be true" (p. 241).

This work appeared in 1972 in the first edition (in Russian; Moscow: Nauka), and then somewhat revised and expanded in 1975, edited by H. Wessel (in German; Berlin: Akademie-Verlag). We are grateful to Dr. Wessel for his encouragement and advice in the preparation of this English-language edition. We are also very glad to recognize the careful and intelligent translation into English that has been provided by Dr. Olga Germogenova of Moscow, who consulted with the author throughout her work; we must also thank A. A. Zinov'ev for his patient cooperation through many years and in the course of his personal difficulties in Moscow and Munich, and not least for the extensive revisions and expansions which he prepared for this English translation which is the work's third edition. We also wish to note Zinov'ev's related paper on 'The Non-traditional Theory of Quantifiers', in *Language, Logic, and Methods* (Boston Studies in the Philosophy of Science, Vol. 31, 1983, pp. 355–408).

We thank Carolyn Fawcett for her editorial assistance throughout our final work on the manuscript, and Katie Platt for her careful proof correcting.

Center for the Philosophy and History of Science
Boston University
May 1983

ROBERT S. COHEN
MARX W. WARTOFSKY

PREFACE

The present monograph contains a brief and, as far as possible, popular discussion of a logical conception developed by the author over a period of many years; its fundamentals are given in a number of works, mainly in four books: *Philosophical Problems of Many-Valued Logic* (Translation from the Russian; Humanities Press, New York; and Reidel, Dordrecht, Holland, 1963), *Foundations of the Logical Theory of Scientific Knowledge* (Complex Logic) Revised (Translation from the Russian [*Boston Studies in the Philosophy of Science*, Vol. 9], Reidel, Dordrecht, Holland, 1973), *Komplexe Logik* (Translation from the Russian; Berlin, Braunschweig, Basel, 1970), *Logical Physics* (Moscow, 1972). The author wishes to note that the formation of this conception is to a considerable degree based on the ideas of such outstanding logicians as J. Łukasiewicz, K. Ajdukiewicz, C. I. Lewis, H. Reichenbach, R. Carnap, G. von Wright. The author also considers it his pleasant duty to thank H. Wessel, A. A. Ivin, G. A. Smirnov, A. M. Fedina, E. A. Sidorenko, G. A. Schegol'kova, G. A. Kuznetsov and V. Shtel'tsner, who were the author's collaborators for some time and promoted the formulation and development of a number of ideas. The author is especially grateful to H. Wessel whose contribution was of substantial importance.

The task of the logical orientation, presented here, is seen by us first of all as that of formulating a sufficiently accurate and rich language for the methodology of experimental sciences, in terms of which a rigorous deduction might be realized. It should be a special language. And as every special language, it must not be translated into ordinary language for the simple reason that it represents just an addition to ordinary language and a development of the latter in a certain direction. The special language would have been superfluous if it were possible to translate it into the ordinary one. And when, in further discussion, we shall formulate certain statements written in the language of logic, also in terms of ordinary language, our purpose will be exclusively that of explanation.

The language which is at present used for discussing the problems of the methodology of experimental sciences represents a set of ill-defined, ambiguous, unstable and logically unrelated linguistic expressions. In addition, they do not provide an adequate and forthright description of the methodological situation in modern science. As far as we understand it, logic does not imply the improvement of this existing language. It implies only the elaboration of logic in this direction in general, an elaboration based on deeper foundations; the final result of such an elaboration is supposed to be a set of linguistic means that can be evaluated only in a retrospective way as an improvement of the language of scientific methodology.

Our assertions, accepted in logic and obtained in it as corollaries, do not represent the results of generalization of experimental material, nor can they be interpreted as assumptions with respect to the objects of reality. They are nothing but parts of the definitions of linguistic expressions or are obtained from such definitions as logical consequences.

Let us give an example. The question whether a physical body can simultaneously occupy different places is usually answered in the negative. And the question, why it is impossible, is usually answered: this is how the world is made. But the structure of the world has nothing to do with it. And how can one guarantee that our assertion will be valid at all past and future times and in all spatial locations? Our confidence in the fact that a physical body cannot simultaneously occupy different places is a logical consequence of the implicit definition of the expression 'different places'. Indeed, when are the locations (spatial domains) regarded as different? From the intuitive point of view two locations x and y are different if and only if they do not have common points. But real 'points' are physical bodies. So if one makes the definition of the expression 'different locations' explicit, the following is obtained. Two locations x and y are considered (called) different if and only if for any physical body a the following assertion is valid: if a occupies one of x and y, it does not occupy at the same time the other one of them. From this definition an assertion logically follows: a physical body cannot simultaneously occupy different places.

We do not have to stress the importance of the analysis of the above linguistic expressions. Complaints about the ambiguity and uncertainty of terminology became commonplace. Monstrous forms and dimensions are acquired by speculations related to the obscurity of terminology. Even comparatively simple problems turn out to be practically unsolvable due to the ignorance or neglect of the logical methods of constructing terminology. Consider the following interesting example. It appears obvious that a process having no beginning does not start. And if a process does not start, it does not exist. But the Universe is a process having no beginning in time. And at the same time the Universe exists. How can these statements be reconciled? We leave it to the reader to resolve this paradox, and if she succeeds, let her take notice of the role belonging to terminological aspects in this situation. The work necessary for solving this problem would be just a special case of the work that is supposed to be raised to the professional level by logic.

We do not consider it to be our task to review the state and history of this kind of logical research. It is our intention to offer to the attention of the reader a special point of view concerning a certain set of problems; our presentation will be based only on well-known and popular facts of the linguistic practice of humans.

To understand this book, no special training in logic is required. What is required is just a sincere desire to investigate the essence of the matter, some patience and tolerance. Although the monograph represents further development of the ideas discussed in the author's works cited above, and an addition to them, it can be understood independently of them as a completely autonomous construction.

A. A. Zinov'ev

CHAPTER ONE

THE GENERAL THEORY OF INFERENCE AND TERMS

1. The Rules of Logic

Logic deals with terms and statements, in other words, with certain phenomena of language. Examples of terms: 'table', 'cow', 'atom', 'Tambov', 'negatively charged particle', 'the fact that all even numbers can be divided by two without a remainder', 'moving with a velocity of 10 km/hr', etc. Examples of statements: "The electron is charged negatively", "All even numbers can be divided by two without a remainder", "If an electric current is passed through a conductor, a magnetic field is produced around it", etc.

By means of special linguistic means (term-producing and statement-producing operators) one can form from one group of terms and statements, new terms and statements. These special means are such words as 'and', 'or', 'not', 'all', 'some', 'if, then', 'which', etc. A study of terms and statements at the same time represents a study of these operators, exactly in the same way as the description of the properties of the latter represents the description of the properties of those terms and statements that contain them.

Logic formulates certain rules for manipulating terms and statements, rather than just discovering them in ready-made form in existing linguistic practice. In the linguistic practice of humans, the above rules appear on an elemental level with all of the associated consequences: vagueness and lack of clarity, variability depending on the specific context, combined with instability, a strong association with the concrete material, their being fragmentary, and so on.

Logic takes elemental human habits into consideration with respect to manipulating terms and statements, and continues this inventive activity of mankind. But this continuation is at a professional level, with the necessary rigour and system. Moreover, in many cases of elemental linguistic practice the above rules do not develop in general, and logic has to introduce them literally for the first time. And scientific development does not change this situation in principle. By analogy with the fact that the development of science does not automatically mean the development of one or another section of mathematics, it does not automatically produce the corresponding rules of logic. The scientific evolution can only provide a stimulus for the development of logic, and to some extent determine its circle of problems. But the logical rules themselves should be formulated by experts in logic, according to the laws of this profession. An opinion to the effect that logical science tells people about facts that are already known to them in their linguistic and cognitive practice is a prejudice.

Logic establishes such rules for terms and statements, which, according to the very method of their formulation, are independent of the specific sphere of linguistic usage, of the particular properties of one or another language as well as of those beings and devices that operate with language. The only difference that is possible here is that certain rules of logic are used in one case and not in others; that sometimes one kind of logical rule is used, and in other situations, a different one, etc. But this difference is as common and trivial as is the difference in the usage of a tablespoon or a synchrotron in other spheres of human activity.

There exists a widely spread opinion that logic can be applied as a special research apparatus to the very object field studied by one or another science (to relay-contact networks, to systems of brain cells, etc.) This opinion is also a prejudice. The sphere of application of logic is language and only language. Besides, one can speak of the application of logic here only in a certain metaphorical sense. The study of logic may affect the behavior of humans in certain linguistic situations. Logic can participate in the process of improvement of language. The effect of logic on the existence of mankind as a rule remains unnoticeable for the outside observer and has nothing to do with those sensational 'applications of logic' that were so much talked about during the last decades.

2. Terms

The terms may be represented by individual words, groups of words, letters, symbols and their combinations. It is an extralogical question, which fragments in one or another specific language are considered as terms. In order to apply the rules of logic, one should possess a certain practical skill for decomposition of the speech flow into those elements for which the rules are formulated. For instance, in order to apply to the expression 'All metals are electrically conducting' the rules of the logical theory of quantifiers and predication, one has to be able to identify in this expression the logical operator (quantifier) 'all', the subject-term 'metal', the predicate-term 'electrically conducting' and the predicativity operator connecting these terms into a whole statement. This predicativity operator is expressed in this case through writing the terms in a sequence. The rules of logic are applicable to terms of specific languages only according to the following scheme: if a given phenomenon is a term, such and such assertions of logic hold for it. A similar principle is valid for statements.

Each term denotes or names certain objects (the word 'object' here can mean anything), i.e., it has a value. Thus, the term 'table' denotes tables; the essence of the value of this term is that it denotes exactly tables, and not other objects; precisely which objects are denoted by the term 'table' can be established by pointing to individual visible tables, by means of demonstrations of drawings or photographs of tables, through description of tables by a set of words and sentences; in physics the term 'elementary particle' denotes physical objects, the description and enumeration of which can be found in the corresponding works on physics.

We shall say that a term b is subordinated by value to a term a (or that an object a is an object b; or, to put it more briefly, that a is b) if and only if the following is valid: any object denoted by the term a can be denoted also by the term b. For instance, a term 'number' is subordinated by value to a term 'an even number' (or, an even number is a number); a term 'elementary particle' is subordinated by value to a term 'electron' (or, an electron is an elementary particle).

If a is b (or a term b is subordinated by value to a term a), in symbolic form it will be written as

$$a \rightharpoonup b.$$

If a is not b (i.e., not every object denoted by the term a can be also denoted by the term b), the symbolic form of it will be

$$\sim(a \rightharpoonup b).$$

If $a \rightharpoonup b$ and $b \rightharpoonup a$, we shall say that the terms a and b are equal in value and write it symbolically as

$$a \rightleftharpoons b.$$

Examples of terms, identical in value: 'diamond' and 'equilateral quadrangle'; 'table', 'der Tisch' and 'стол'.

The relation of subordination of terms by value may be used to define other relations of terms, in particular, the following one: a term a is called common (generic) with respect to a term b, and a term b is called special (a species) with respect to a if and only if $b \rightharpoonup a$ and $\sim (a \rightharpoonup b)$. For example, the term 'diamond' is a species term with respect to the term 'quadrangle', while the latter is generic with respect to the former.

3. Two Types of Terms

Consider a statement 'The electron is negatively charged'. It consists of the terms 'electron' and 'negatively charged'. But the roles of those terms in the statement are different: the first denotes the object under consideration, whereas the second denotes what we would like to say about this object. The terms of the second type are called predicates, and what is denoted by them is called the attributes of objects. The terms of the first type are called subjects.

Predicates are divided into one-, two-, three-, etc. place ones, depending on how many term-subjects are required to form a statement. Thus, a predicate 'is negatively charged' is one-place, and the predicates in statements 'a is greater than b' and 'a is between b and c' are, respectively, two-place and three-place. We assume that it is possible to distinguish between predicates in such a way. In exactly the same manner we suppose that it is possible to distinguish between predicates and subjects. Sometimes, as in the example given at the beginning of this paragraph, the difference is obvious, in other cases a certain effort is required to discover and single out the predicate. Thus, the predicate in the statement 'a is greater than b' is the expression 'the first object is greater than the second object', which cannot be visualized in it directly.

No predicate represents a subject and no subject represents a predicate. For any a and b, if one of them is a subject and the other is a predicate, the following is valid: $\sim (a \rightharpoonup b)$ and $\sim (b \rightharpoonup a)$. In order to 'transform' the subject a into the predicate b, or vice versa, one should construct of a (or of b) and other additional linguistic elements the term b (or, respectively, a) according to special rules.

4. Simple and Complex Terms

From given terms and statements one can form the new terms by applying certain special means that we shall call term-producing operators, or, in abbreviated form, T-operators. For example, from the terms 'elementary particle' and 'negatively charged' a complex term 'elementary particle which is negatively charged' may be produced. The word 'which' here plays the role of the T-operator. From the statement 'The electron is negatively charged' one may form a term 'The fact that the electron is negatively charged' by applying the T-operator 'The fact that'.

Those terms that cannot be decomposed into other terms or statements, and T-operators, will be called *logically simple*. On the other hand, those terms that contain other terms or statements and T-operators, we shall call *logically complex*.

It is not within the sphere of competence of logic to judge what combinations of words and letters are regarded as simple terms and what as complex. Thus, a question whether the term 'elementary particle' is simple or complex is an extralogical question. Obviously, this term may be regarded as complex if it is interpreted as the term 'a particle, which is elementary'. But it can be also considered as simple, if it is introduced in a different way, as a whole.

5. Complex Terms

One of the ways of forming complex terms is the creation of pairs, triplets, and, in general, *n*-tuples of terms. We shall write them symbolically as $(a,b), (a,b,c), \ldots, (a^1, \ldots, a^n)$, where $n \geqslant 2$. The location of these terms in statements usually differs from that given above. For instance, in the expression 'a is greater than b' the term representing a pair of terms (a, b), is distributed over different parts of the statement. Suggesting the symbolic form $(a^1, \ldots, a^n)$, we realize the abstraction or schematize the real situation. But this abstraction is justified if only for the reason that any statement x, containing the subject-terms $a^1, \ldots, a^n$, may be written in the form of a statement with the term $(a^1, \ldots, a^n)$ without any loss of information (in modern language). For example, the expression 'a is between b and c' may be given the form '(a,b,c) are such that the first of them is between the second and the third'. The objects, denoted by this kind of terms, we shall call pairs, triplets, etc., or in general, *n*-tuples of objects. The symbol $(a^1, \ldots, a^n)$ may be read also as '*n*-tuple of objects $a^1, \ldots, a^n$'.

Let a be a subject, Pa predicate, xa statement. They may be used to form complex terms 'a, which has an attribute P', 'a such that P', 'a such that x is true', 'a such that x', 'a, with respect to which x holds', etc. Here we have employed the T-operator that will be called the *operator of limitation* (or the operator 'which', 'such, that') and designated by a symbol $\downarrow$. The terms with this operator we shall write through symbols such as

$a{\downarrow}P$, $a{\downarrow}x$, $(a^1,\ldots,a^n){\downarrow}x$ and so on. Examples of these terms: 'A particle which is negatively charged', 'A pair of numbers such that one number is greater than the other', 'A number a such that the number a is a prime'. The limitation operator also participates in forming terms with negations. For instance, 'A particle which is not negatively charged'.

6. Occurrence of Terms and Statements in Other Terms and Statements

Terms and statements can constitute parts of other terms and statements, assuming the role of terms and statements or that of physical bodies (sounds, lines on paper) of a certain kind. In the first case we shall speak of occurrence as a term or a statement, in the second, of the physical occurrence. For example, in the expression 'A statement x is true' the statement x occurs physically, while in the expression 'If x, then y' the proposition x occurs as a statement. In the expression 'A term "table" is a subject' the word 'table' occurs physically, and in the expression 'The table is wooden', as a term. Accurate discrimination between these cases is done in logic through enumeration of the structures of linguistic expressions, in which the terms and statements occur exactly as such. On the other hand, the general principle is the following: if, according to the accepted rules of the formation term, from the assumption to the effect that $a^1,\ldots,a^n$ $(n \geqslant 1)$ are terms (or statements), and $b^1,\ldots,b^m$ $(m \geqslant 1)$ are logical operators, it follows that an expression A, formed of $a^1,\ldots,a^n$, $b^1,\ldots,b^m$ according to these rules, is a term (or a statement), then the terms (statements) $a^1,\ldots,a^n$ as well as A itself occur in A as terms (or statements).

On the basis of the above, we can accept the following rule: if $a \rightleftharpoons b$, and a term d is formed of a term c through replacement of one or more occurrences of a in c as a term by b, then $c \rightharpoonup d$. It follows from this rule (under the same limitations for the occurrence) that if $a \rightleftharpoons b$, then $c \rightleftharpoons d$.

7. Metaterms and Metastatements

The terms denoting terms and statements are metaterms. The statements containing metaterms are metastatements. For instance, the expression 'A term "table"' is a metaterm with respect to the term 'table', while the statement 'The term "table" is simple' is a metastatement.

Every metaterm is a subject. No metapredicates exist. If *a* is a predicate, and *b* is a term denoting it, *b* is still a subject.

Of special importance are the metaterms for the formation of which the designated terms and statements themselves are responsible, including such terms as 'a term *a*', 'a subject *a*', 'a predicate *a*', 'a statement *a*'. Our generalized and abbreviated notation for them will be m*a*, where *a* is the denoted term or the statement. The following negative rules hold for such terms:

(1) $\sim(a \rightarrow \mathrm{m}a) \vee \sim(\mathrm{m}a \rightarrow a)$,

(2) $\mathrm{m}a \rightarrow \mathrm{m}b$ does not follow from $a \rightarrow b$,

(3) $\mathrm{m}a \rightleftharpoons \mathrm{m}b$ does not follow from $a \rightleftharpoons b$.

One may get a wrong impression, as if *a* occurs in m*a* as a term or a statement. But this is not so. In reality, *a* occurs in m*a* as a set of physical objects (sounds, lines, etc.) that constitute the matter of language. For a greater degree of clarity the expression m*a* should be read as 'A term (a subject, a predicate, a statement), which is written, pronounced, etc. as *a*'. The specific feature of m*a* is that it really contains the denoted term or the statement *a*; this term or statement, being a component of m*a*, already does not represent a term or a statement. The term or statement *a* does not ocur in m*a* as a term or, respectively, as a statement.

If one assumes that *a* occurs in m*a* as a term, the result is the following. If $a \rightleftharpoons b$, then, according to the rule of substitution of Section 6, we obtain that $\mathrm{m}a \rightleftharpoons \mathrm{m}b$. But if the external appearance (physical) of *a* and *b* is different, then m*a* and m*b* will denote different objects; m*a* cannot denote *b*, and m*b* cannot denote *a*. Therefore, $\mathrm{m}a \rightleftharpoons \mathrm{m}b$ will be wrong. In order to avoid such paradoxes, one should define explicitly these cases where the terms and statements appear in the linguistic expressions as terms and statements, and these cases where they occur as physical objects (in particular as it was done above).

Consider a proposition 'N states that the writer Leo Tolstoi is the author of the novel "Anna Karenina"'. The proposition 'N states that the writer Leo Tolstoi is a writer Leo Tolstoi' does not follow from it logically, although the term 'a writer Leo Tolstoi' is identical (according to the assumption) to the term 'the author of the novel "Anna Karenina"'. The point is that the initial proposition has the following logical structure. Its subjects are the terms 'N' and a statement 'The writer Leo Tolstoi is the author of the novel "Anna Karenina"'. As one may see, the second term is a term of the kind m*a*. Its predicate is the expression 'The first asserts the second'. So that the expression 'The writer Leo Tolstoi is the author of the novel "Anna Karenina"' does not occur in the statement under consideration as a statement, while the terms 'The statement "The writer Leo Tolstoi is the author of the novel "Anna Karenina""' and 'The statement "The writer Leo Tolstoi is a writer Leo Tolstoi"' are not identical as far as meaning is concerned.

Let us note incidentally that the above expressions $a \rightharpoonup b$ and $a \rightleftharpoons b$ cannot be regarded as consisting of subject-terms a and b and predicates $\rightharpoonup$ and $\rightleftharpoons$. They represent expressions consisting of subject-terms ma and mb and predicates $\rightharpoonup$ and $\rightleftharpoons$.

In general, the laws of logic are metastatements whose subjects are the metaterms ma, where a is a term or a statement, and the predicates are $\vdash$, $\rightharpoonup$, 'true', 'false', 'provable', etc. Logic represents a metascience with respect to all other sciences and, incidentally, the only metascience at all, if one considers it to be a necessary attribute of science to include a generalized study of one or another sphere of objects. So that the logical laws are not statements of the form $x \vee \sim x$, $x \wedge y \supset x$, etc., which is the common conviction, but either metastatements of the type $\vdash (x \vee \sim x)$, $x \wedge y \vdash x$, etc., or metastatements such as 'm$(x \vee \sim x)$ is a tautology if certain truth tables for $\vee$ and $\sim$ are accepted, together with such and such definition of tautology', 'From m$(x \wedge y)$ follows mx', etc.

Let us make one more remark that the reader may return to after going through Sections 12 and 15. For any predicate P the meanings of P(a) and P(ma) are established independently of each other. There is also no logical relation between the truth values of P(a) and P(ma). For instance, an expression 'Phlogiston exists' is not true while the expression 'The term "phlogiston" exists' is true. However, there is this kind of dependence here: the meaning of ma is known if and only if one knows the meaning of a; and if a is not a term, then ma is not a term.

8. The Meaning of Terms

We shall say that the investigator (i.e., the person who is working with terms) knows the meaning of the terms if and only if she knows the value of all terms occurring in a given term, and also knows the properties of all logical operators occurring in this term.

For example, we know the meaning of the term 'a number that can be divided by two without a remainder' if and only if we know the value of the terms 'number' and 'can be divided by two without a remainder', as well as the properties of the operator 'which'. The properties of this operator are defined in logic.

Let us accept an assertion: if an investigator knows the meaning of a term, she knows the value of that term as well.

By means of the above assertion it is possible to establish the value of terms through determining their meaning. For instance, the value of a term 'A round square' can be established (learned) in just one way, namely, by determining the value of the terms 'square' and 'round' and the properties of the operator 'which' (it is not expressed clearly here; in a more explicit form the structure of this term is as follows: 'A square which is round'). A considerable portion of the terms of contemporary science has this property. We would like to underline the fact that if we establish the value of a term through determining its meaning, nothing else is required. If, for instance, when answering the question, what is the value of the term 'A round square', we were to say in addition to explaining the value of terms 'round' and 'square' and the property of the operator 'which', something different, we would commit a logical error. In such cases it is necessary to stop in time. When the search for the value of a term is overdone, it is no better than when it is insufficient.

9. Terms Including Statements

One can form the subject-terms out of statements according to the following rule: if x is a statement, then the expression 'The fact that x' is a term, and at the same time, a subject. In this case the expression 'The fact that' appears in the role of a T-operator. This kind of term will be denoted by such symbols as $\downarrow x$, where x is a statement and $\downarrow$ is a T-operator 'The fact, that'.

One can also form the predicate-terms out of statements by using the following rule: if x is a statement, then the expression 'Such that x' ('is characterized by the fact that x') represents a term (a predicate). These terms will be denoted by such symbols as $x\downarrow$ where $\downarrow$ is a T-operator 'Such that'.

10. Definitions

Definitions of terms may be interpreted as only one of the ways of the introduction of terms into the language. But this is the most important way.

Consider a simple definition: 'A diamond is an equilateral tetragon'. It may also be written in a different form: 'An equilateral tetragon is called a diamond' or 'We shall call by a diamond an equilateral tetragon'. But the following is invariant during all the variations: (1) by virtue of the definition we introduce a new term 'diamond'; (2) a decision is taken to consider the word 'diamond' as a term, equivalent to the term 'equilateral tetragon'.

Thus, if the logical structure of this definition is totally elucidated, it will take on the following form: an expression of the type 'diamond' (the word 'diamond') will be considered a term such that 'diamond $\rightleftarrows$ 'equilateral quadrangle'. The term 'diamond' is here called the defined, while the term 'equilateral tetragon', the defining. The second one is a complex term of the form $a\downarrow \mathrm{P}$ ('a tetragon, which is equilateral', or 'a tetragon, all sides of which are equal'). Its meaning is regarded as known. The meaning of the first term (the defined) is known only by virtue of the fact that it is considered equivalent to the second one (the defining).

Above we have discussed an example of definitions that are constructed according to the following rule: 'An object a will be considered a term such that $a \rightleftarrows b$, where b is a term'. In abbreviated form,

$$a = \mathrm{Df.}\ b.$$

Another formulation of this definition is as follows: according to our decision (by definition), a is a term, whose meaning is identical with that of a term b. Such definitions are useful in practice only when b represents a complex term, and a is either a simple term or contains the newly introduced simple term.

The above definitions may be classified into two types: (1) a is a simple term; (2) a is a complex term, containing a simple one. For instance, an expression 'a provable formula' contains a term 'formula' the value of which is known, and a newly introduced term 'provable' which is defined with respect to formulas.

There are various forms of definitions. A definition that occurs frequently, and is based on the specification of the genus and species distinction, has the following structure:

$$a = \mathrm{Df.}\ b\downarrow \mathrm{P},$$
$$a = \mathrm{Df.}\ b\downarrow x,$$

where P is the species attribute of certain objects b, and x is a statement concerning the species attribute. The definition of a through $b\downarrow x$ is not necessarily of the genus-species type. Consider the following definition: the term *a* will denote the objects *b* such that *b* are not objects *c* and *d*. There is no indication of any species attribute P here. The only thing that we encounter here is just the exclusion: *a* are all *b* except *c* and *d*.

The definition through enumeration of the species of *a* is constructed according to the following rule: let *a* be a term such that b^1 is $a, \ldots, b^n$ is *a*, and no other object is *a* (or that $b^1, \ldots, b^n$, and only these objects, are *a*).

If the definition $a = \text{Df. } b$ is adopted, then one accepts the assertion $a \rightleftharpoons b$. If one adopts the definition through enumeration of the species of *a*, the following set of assertions is accepted: (1) $b^1 \rightharpoonup a, \ldots, b^n \rightharpoonup a$; (2) if $\sim(c \rightharpoonup b^1), \ldots, \sim(c \rightharpoonup b^n)$, then $\sim(c \rightharpoonup a)$, where *c* neither contains *a* nor is defined through it.

The recursive definitions represent a more complex form of the definition through enumeration. The simplest of them are constructed according to the following rule. Let *a* be a term such that (1) $b^1 \rightharpoonup a, \ldots, b^n \rightharpoonup a$ (where $n \geqslant 1$); (2) if $c^1 \rightharpoonup a, \ldots, c^m \rightharpoonup a$ (where $m \geqslant 1$), then $d^1 \rightharpoonup a, \ldots, d^k \rightharpoonup a$ (where $k \geqslant 1$); (3) no other object, except as specified in (1) and (2), is *a*.

Further we shall consider in greater detail some other forms of definition that play a primary role in our discussion. In general, the main task of the present work is the definition of linguistic expressions. But in order to discuss this topic at a more rigorous and detailed level, we should for the time being consider statements.

11. Statements

Statements represent special linguistic constructions, formed of terms, statements, and statement-producing operators (B-operators).

The simplest statements consist of a subject, a predicate, and a predicativity operator, which connects the subject and the predicate into a whole statement. The examples of such statements are: 'The electron is negatively charged', '7 is greater than 5', 'The number 5 is a prime', etc. These statements will be denoted by symbols such as P(*a*) and $P(a^1, \ldots, a^n)$, where $n \geqslant 2$; the above symbols will be read as 'A subject *a* has an attribute P' and 'Subjects $a^1, \ldots, a^n$ are such that P' (or, 'An *n*-tuple of objects $a^1, \ldots, a^n$ has an attribute P'). Certainly, this is a schematization. The reader is supposed to possess the skill of transforming statements of specific languages to such form.

Let us list the B-operators, by means of which from given statements (in the final analysis, from the simple ones) one can form new statements:

(1) $\wedge$ – conjunction ('and'; 'each of');
(2) $\vee$ – weakened disjunction ('or'; 'at least one of');
(3) : – strong disjunction ('either, or'; 'one and only one of');
(4) $\sim$ – external negation ('not'; 'not so');
(5) $\rightarrow$ – the conditionality operator ('if, then');
(6) $\leftrightarrow$ – the operator of reversible conditionality ('if and only if');
(7) $\forall$ – the universal quantifier ('all');
(8) $\exists$ – the existence quantifier ('certain');
(9) $\urcorner$ – internal negation (it is read in the same way as external negation, but its location in the statements is different);
(10) ? – the uncertainty operator (its meaning will be explained below).

We shall illustrate the relation of these operators and the respective linguistic means with the example of conjunction. First, logic uses as conjunction, in addition to $\wedge$, various symbols, in particular, a point or dot, &, K, etc. Second, in various languages and in various cases of the same language the role of $\wedge$ [i.e. conjunction] may belong not only to the word 'and', but also to other means, in particular, 'but', a comma, a semicolon, 'and also', 'besides', etc. And finally, the above linguistic means, in addition to sometimes performing the functions of conjunction, carry out other functions as well. So that logic, introducing the conjunction operator abstracts just a certain property from various linguistic means and treats it as such – a method common to logic and other sciences.

Statements with operators (1)–(10) have the following logical structure:

(1) $x \wedge y$ – 'x and y'
$(x^1 \wedge x^2 \wedge \ldots \wedge x^n)$ – 'x^1 and x^2 and ... and x^n' ('Each of $x^1, x^2, \ldots, x^n$');
(2) $x \vee y$ – 'x or y';
$(x^1 \vee x^2 \vee \ldots \vee x^n)$ – 'x^1 or x^2 or ... or x^n' ('At least one of $x^1, x^2, \ldots, x^n$');
(3) $x : y$ – 'Either x or y';
$(x^1 : x^2 ; \ldots : x^n)$ – 'One and only one of $x^1, x^2, \ldots, x^n$';
(4) $\sim x$ – 'Not x' ('Not so as is stated by x');
(5) $x \rightarrow y$ – 'If x, then y';
(6) $x \leftrightarrow y$ – 'x, if and only if y' (or 'If x, then y; if y, then x');
(7) $(\forall a)x$ – 'All a are such that x' ('All a x'; 'For all a, x');
(8) $(\exists a)x$ – 'Certain a are such that x';
(9) $\urcorner P(a)$ – 'The object a has no attribute P' ('a does not have P');

$(\neg\forall a)x$ – 'Not all a are such that x';
$(\neg\exists a)x$ – 'there are no a such that x';

(10) $?P(a)$ – 'It is impossible to say whether $P(a)$ or $\neg P(a)$' ('Not – $P(a)$ and not – $\neg P(a)$'; 'It is not known, whether $P(a)$ or $\neg P(a)$');
$(?\forall a)x$ – 'Not – $(\forall a)x$ and not – $(\neg\forall a)x$'; 'It is impossible to establish whether $(\forall a)x$ or $(\neg\forall a)x$';
$(?\exists a)x$ – 'Not – $(\exists a)x$ and not – $(\neg\exists a)x$'; 'It is impossible to establish whether $(\exists a)x$ or $(\neg\exists a)x$'.

The structures of the statements listed above do not exhaust all possible forms of statements. Other kinds of statements are formed by means of introduction of the derivate B-operators, by means of making the structure of the terms more complex and through redistribution of their parts in the linguistic constructions. For example, the statement $P(a)$, where a is a term 'b under condition c' may be transformed into a statement '$P(b)$ under condition c', in which the reference to conditions is taken out of the subject as a special part of the statement.

12. The Meaning of Statements

We shall say that the meaning of a statement is known if and only if one knows the meaning of all terms occurring in it, as well as the properties of all its logical operators.

The meaning of two statements x and y will be considered identical if and only if $\downarrow x \rightleftharpoons \downarrow y$. The identity of meaning will be denoted by symbols such as $x \equiv y$.

Obviously, if $x \equiv y$ and $y \equiv z$, then $x \equiv z$. By virtue of the rule of substitution of Section 7, the following is valid: if $x \equiv y$, and at the same time v is formed of z as a statement on y, then $z \equiv v$.

13. Definitions with Statements

Terms may be introduced in such a way that the whole statement is immediately defined, in other words, terms may be introduced as parts of statements. These definitions are constructed according to the following principle: 'We shall consider x to be a statement such that $x \equiv y$', where y is a given statement or a set of statements. In symbolic form this definition will be written as

$$x \equiv \text{Df. } y,$$

where x is the defined statement, containing the defined term, and y is a given statement through which x is defined (i.e., the defining statement).

If one adopts a definition $x \equiv \text{Df. } y$, this means the adoption of the statement $x \equiv y$.

More frequently (for the sake of convenience) this kind of definition is written in the following form: 'We shall say that x if and only if y'. For instance, 'We shall say that a formula is provable in S if and only if it is an axiom of S or is obtained from the axioms in S according to the rules of inference in S'. This form simply implies elimination of a part of the means of definition (for example, reference to the meaning of statements); this makes the inference of the consequences from the definition easier. But in this case the nature of the definition is not explicitly stated, which has certain shortcomings.

By means of the above definitions, the predicates and logical operators are defined. For instance, the operators of strong disjunction, reversible conditionality and uncertainty may be defined through other operators by the following method:

$$(x : y) \equiv \text{Df.} (x \vee y) \wedge (\sim x \vee \sim y),$$
$$(x \leftrightarrow y) \equiv \text{Df.} (x \rightarrow y) \wedge (y \rightarrow x),$$
$$?\text{P}(a) \equiv \text{Df.} \sim \text{P}(a) \wedge \sim \urcorner \text{P}(a),$$
$$(?\forall a)x \equiv \text{Df.} \sim (\forall a)x \wedge \sim (\urcorner \forall a)x,$$
$$(?\exists a)x \equiv \text{Df.} \sim (\exists a)x \wedge \sim (\urcorner \exists a)x.$$

Below we shall give numerous examples of the definitions of predicates. For the meantime, we shall limit ourselves to a somewhat abstract, but a more visually obvious example:

$$\text{P}(a) \equiv \text{Df. Q}^1(a) \wedge \text{Q}^2(a) \wedge \urcorner \text{Q}^3(a),$$

where Q^1, Q^2, Q^3 are certain attributes, and the predicate P is introduced for the sake of abbreviation.

Such definitions as $a = \text{Df. } b$ allow us to replace every occurrence of b as a term by a, and vice versa. On the other hand, the definitions of the kind $x = \text{Df. } y$ allow us to replace the occurrence of x as a statement by y, and vice versa. If in this process the replacement of terms takes place, it is only a consequence of the replacement of statements.

Definitions of both of the above kinds provide an opportunity for us to replace certain linguistic complexes by those that are in a sense more convenient, (for instance, by more compact ones, by those more intuitively obvious, etc.). E.g., the word 'diamond' is both shorter and more convenient than the combination of words 'equilateral tetragon'; the expression $x:y$ is shorter than $(x \vee y) \wedge (\sim x \vee \sim y)$. Starting from a certain moment, in the absence of such abbreviations and replacements, the record of knowledge and operating with it become practically impossible. The search for the most convenient forms of abbreviation represents one of the most important tasks in the construction of scientific language in general.

14. The Definition of Predicates

The definition of predicates deserves special attention. The predicates are defined as parts of complex terms or as parts of statements. In order to indicate in some way which attribute is denoted by a given predicate P, one should indicate certain objects possessing this attribute. The attributes of objects are the attributes of definite objects and cannot exist in separation from them. So that if a predicate P is introduced through a definition, the definition is concerned not with P by itself, but with such terms as a $\downarrow$P, $\downarrow$P(a), etc., or with a statement x, containing P.

This peculiar feature of predicates leads to another one. The predicates are defined for a more or less broad circle of specified subjects (objects), and not always for any subjects. If, for example, P is defined through defining $a \downarrow$P or $\downarrow$P(a), then the sphere of applicability of P depends entirely on the degree of generality of a. If $b \rightharpoonup a$, then P is applicable to b, the expression $b \downarrow$P is a term, and P(b) is a statement. If, on the other hand, $\sim(b \rightharpoonup a)$, and there is no such term c for which the construction of the definition of the term $c \downarrow$P or the statement P(c) may be carried out, and for which $b \rightharpoonup c$, then $b \downarrow$P does not represent a term, while P(b) is not a statement. These are meaningless linguistic expressions that just resemble terms and statements. And any expression, containing them with the claim of the occurrence of a term and a statement, represents neither a term nor a statement.

To give an example, the predicate 'even' is defined only for the terms of integers and is not defined, say, for the terms of metals. The expression 'Even copper' is not a term at all. Similarly, the expression 'yellow integral' is not a term either, while the expressions 'An integral is yellow' and 'An integral is not yellow' do not represent statements. And it is impossible to say that they are not true, since the predicate 'not true' is defined only for those expressions that are statements.

It is only when a is an 'object' term that the definition of term $a \downarrow$P or the statement P(a) represents a definition of P for any objects, since for such a the following is valid: for any term b, if b is a subject, we have $b \rightharpoonup a$.

15. The Truth Values of Statements

The evaluation of statements as true and false is well known. It is now supplemented by uncertainty, non-verifiability, undecidability, etc. The expressions 'true', 'false', 'uncertain', and so on are considered to be terms (or symbols) of statements' truth values. Below we shall give a formulation of the fundamental principles concerning the truth values of statements.

First of all, the term 'true' should be defined as a predicate of such statements as 'The expression x is true'.

Second, depending on the differences in the structure of the statements x, the definition of the term 'true' should be different. For instance, for propositions $x \wedge y$ with the operator $\wedge$ and propositions $x \vee y$ with the operator $\vee$, the term 'true' is defined differently (otherwise these operators will not be differentiated). Regardless of the structure of x, only the following definition is appropriate: 'x is true' $\equiv$ Df. x ('x is true if and only if x'). This definition provides the rule for introduction of the predicate 'true' into the language as well as for elimination of it as redundant.

Third, for simple statements the term 'true' is accepted without definition. In this case it would be sufficient to give just an explanation and the rule mentioned in the second paragraph. After the term 'true' is accepted for simple statements, it can then be defined for complex statements through truth terms for simple statements. But the definitions should be formulated in such a way that the rule described in the second paragraph should always be valid.

Fourth, all other truth values are defined through the value 'true'. For instance, the value 'not-true' is defined in the following manner: 'x is not-true' $\equiv$ Df. 'x is not true' or 'x is not-true' $\equiv$ Df. '$\sim x$ is true'. Another example: the definitions of the values 'indefinite' and 'false' for statements of the form $\mathrm{P}(a)$: (1) '$\mathrm{P}(a)$ is false' $\equiv$ Df. '$\urcorner \mathrm{P}(a)$ is true'; (2) '$\mathrm{P}(a)$ is indefinite' $\equiv$ Df. '$\sim \mathrm{P}(a) \wedge \sim \urcorner \mathrm{P}(a)$ is true'.

Thus, all the terms of the truth values may be finally eliminated from the language, and the corresponding expressions containing them could be replaced by those with the term 'true'; on the other hand, the latter can be completely removed from the language according to the rule of substituting x in place of the expressions 'x is true'. But this does not mean at all that these terms are in general superfluous. They are convenient for establishing a number of logical operations, and, beginning from a certain moment, they are practically necessary for effective abbreviation of complex linguistic constructions into more compact ones.

While the value 'true' may be eliminated from the language, it cannot be removed from the situation in which the statements are used. It expresses the relation of the statements to the states, expresses the act of acceptance of the statements or of agreeing with what they say. So that the replacement of '*x* is true' with *x* means that the recognition of *x* is realized not through linguistic means (not through the predicate 'true') but in a different way (by the very fact of pronouncing or writing *x*, as is done most frequently).

The reducibility of all truth values to the value 'true' means the following: by asserting that *x* has a certain truth value *a*, we acknowledge the fact that there is a certain state, described by expression *y*, and the definition of *a* should be constructed in such a way that the statement '*x* has a truth value *a* if and only if *y*' may be deducible from it.

Finally, for any truth value the following is valid: any statement either has this truth value or does not have it. In particular, every statement is either true or not true, either false or not false, etc.

The linguistic expression 'truth' is nothing but an abbreviation for the expression 'a statement, which is true'; the situation is similar for other truth values.

16. The Number of Truth Values

Consider a simplest statement P(*a*), for instance 'A particle *a* occupies a space location *b*'. It is possible that: (1) it is impossible to observe the particle *a*; (2) it is possible to observe the particle *a*. In the second case one can imagine the following subcases: (1) the particle *a* indeed occupies the space location *b*; (2) the particle *a* does not occupy the space location *b*; (3) the particle *a* is moving in such a way that we cannot say that it occupies *b*, and at the same time we cannot say that it does not occupy *b*. Thus, for P(*a*) it is possible to introduce four terms of truth values, in correspondence with the four possibilities listed above. If they are indeed introduced, it is done only in order to denote which one of these possibilities is recognized. Assume, for instance, that the following definitions are adopted:

(1) P(*a*) is unverifiable if and only if it is impossible to observe *a*;

(2) P(*a*) is true if and only if we may observe a and indeed *a* has an attribute P;

(3) P(a) is indefinite if and only if it is possible to observe a, but it is impossible to establish whether a has an attribute P or not;

(4) P(a) is false if and only if we may observe a, and a does not possess the attribute P.

And if now we shall say, for instance, that the statement P(a) is false, it would just mean that we acknowledge (recognize) the following: a may be observed, a does not have P.

For statements characterized by more complex structures, there are still greater possibilities. So in principle the number of the truth values is not limited. It depends on practical expediency, how much of them actually appears in the language. Up to the present time the two values 'true' and 'false' were in use, and falsity was understood as the negation of truth (as 'not true'). During recent decades a third value, 'indefinite' became more and more frequently employed. In this connection, the term 'false' proved ambiguous: on the one hand, it began to denote one of the truth values, along with 'true' and 'indefinite', and on the other, it preserved the meaning of the negation of truth. This leads to various kinds of confusion.

As the negation of truth, we shall use the term 'not-true', on the basis of the following definition: a statement is not-true (or represents non-truth) if and only if it is not true. On the other hand, the term 'false' will designate one of the truth values, which only sometimes totally coincides with the value 'not-true', namely, when a statement can take on only one of two values 'true' and 'false'. Thus, if a statement is false, it is not-true; but if a statement is not-true, this does not mean that it is false: it may be unverifiable, indefinite, etc.

Obviously, every statement is either true or not-true. But it is not always (under the condition of the accepted agreement) true that a statement is either true or false: it may be indefinite, i.e., it may be not true and may be not false.

17. The Coordinates of Statements

The truth values of certain statements depend on the place and time of their use. For instance, the expression 'The window is open' may be true with respect to one window and not-true with respect to another (which is closed); it may be true with regard to a certain window at one time and not-true with regard to the same window at a different time. We shall call such expressions local. The set of local statements includes those the truth values of which may vary depending on the individuals involved. But since the spatial and temporal location of various individuals differ in one or another way, the change of the individual to which the statement refers is the change of the spatial and temporal domain of the statement usage. Similarly, one may say that the dependence of the statements' truth values on the conditions can be reduced to spatial and temporal dependence (the change of conditions implies either change of location or change of time).

Let us introduce the term 'statement coordinates' for a statement x to denote the place, time, conditions, etc., for which x is used and its truth value is established. The statement coordinates are usually implicitly present; sometimes they are unimportant, sometimes they are recorded by special symbols. However, when one formulates and applies the rules of logic, it is essential always to assume that in one or another way the statement coordinates may be taken into account. Consider, for instance, an expression 'If $P(a)$, then $Q(a)$'. In order to use it correctly in the case when the statements occurring in it are local, it is necessary to observe the identity of coordinates for $P(a)$ and $Q(a)$, in particular, when such an individual b is chosen so that it coincides with a in the first of them, the same individual should be chosen in the second expression also. If it is given that x is true and y is true, we shall be justified to recognize $x \wedge y$ as true only under the condition of the identity of coordinates for x, y, and $x \wedge y$ (certainly, when the statements x and y are local).

There is no logical relation assumed between the coordinates a and the statement x. The coordinates represent a definite part of a statement, and in order to avoid errors during logical operations with x, this part a should be in some form associated with x, in other words, we should operate with the statement 'x for a'.

Those statements the truth values of which do not vary with the variation of coordinates (which have the same truth value in any coordinates) we shall call universal. Examples of universal statements: 'All even numbers can be divided by two without a remainder', 'The electron is negatively charged', '$5 > 3$', 'Man is mortal', etc.

The rules of logic, corrected for statement coordinates, are equally applicable to universal and to local statements. The former are characteristic of the exact sciences, and since they are dealing with arbitrary coordinates (i.e., indifferent with regard to coordinates), these are in general omitted. The latter are characteristic of the experimental sciences, and frequently coordinate identification is essential for them.

18. Truth Value for Statements with Conjunction, Disjunction and Negation Operators

The predicates 'true' and 'not-true' for statements with operators $\sim$, $\vee$, $\wedge$ are defined as follows:

(1^1) $x \wedge y$ is true if and only if x is true and y is true.
(1^2) $x \wedge y$ is not-true if and only if at least one of x and y is not-true.
(2^1) $x \vee y$ is true if and only if at least one of x and y is true.
(2^2) $x \vee y$ is not-true if and only if x is not-true and y is not-true.
(3^1) $\sim x$ is true if and only if x is not-true.
(3^2) $\sim x$ is not-true if and only if x is true.

The difference between explicit definitions and those given above is that the expression 'if and only if' is replaced by $\equiv$ Df. For the sake of brevity and more clear representation, the implicit definitions are given the form of tables.

It is obvious from the above that the meaning of the term 'true' depends on what kind of statements $x \wedge y, x \vee y, \sim x$, etc. it refers to. And this term has no universal and general meaning, regardless of the statement structure. In general, it has no meaning at all by itself, if taken separately from the expressions where it plays the role of the predicates. The situation for the term 'not-true' is similar.

Moreover, the terms of the truth-values may have different meaning even for expressions with identical structure, depending on the number of truth values involved. Assume, for instance, that there are three truth-values introduced – 'true', 'uncertain' and 'false'. The following definitions of the term 'true' are possible for a statement of the form $\sim x$: (1) $\sim x$ is true if and only if x is false; (2) $\sim x$ is true if and only if x is uncertain or x is false. For the statement $x \wedge y$ the following definitions of the term 'false' are possible: (1) $x \wedge y$ is false if and only if at least one of the expressions x and y is false; (2) $x \wedge y$ is false if and only if one of the expressions x and y is either false or uncertain, and the other one is either false or true. The number of possibilities for such variations increases with increase in the number of truth values.

If the terms of the truth values are considered as given (for instance, fundamentally clear), then the above (and in general, this kind of) definitions may be regarded as definitions of the logical operators $\sim$, $\vee$, $\wedge$. In this case the basic definition scheme has the following form: by definition, operator $\sim$ is characterized by such properties that if x is true (not-true), then $\sim x$ is not-true (true); operator $\wedge$ is characterized by such properties that $x \wedge y$ is true if and only if x is true and y is true; etc. for other cases and for $\vee$.

Depending on the number of truth values, various versions of the operators are possible here. Assume, for example, that the truth values are 'true', 'uncertain', and 'false'. Introduce operators $\wedge^1$, $\wedge^2$, and $\wedge^3$ such that

(1) if x is true and y is uncertain, then $x \wedge^1 y$ is uncertain, $x \wedge^2 y$ is uncertain and $x \wedge^3 y$ is false;

(2) if x is false and y is uncertain or if x is uncertain and y is false, then $x \wedge^1 y$ is false, $x \wedge^2 y$ is uncertain and $x \wedge^3 y$ is false;

(3) if one replaces everywhere in definitions of $\wedge^1$, $\wedge^2$, $\wedge^3$ the truth values 'uncertain' and 'false' by 'not-true', then all these definitions will coincide with the above definition of $\wedge$.

This fact is sometimes interpreted in the sense that in all these cases we are dealing with the same logical operator of conjunction, but the properties of this operator vary depending on the number of the truth values. Such most obvious errors serve as a foundation for pretentious conceptions (in particular, that of a special logic for microphysics). In reality, in all four cases the definitions specify different operators. They can be compared only according to certain criteria. In particular, operator $\wedge^1$, $\wedge^2$, $\wedge^3$ may be regarded as analogs of the two-valued conjunction $\wedge$.

19. Truth Values of Other Forms of Statements

For the expressions P(a) and $\neg$P(a) we accept the term 'true' without definition. But, on the other hand, the term 'not-true' needs an explanation here, since the negation of P(a) is not necessarily $\neg$P(a), while the negation of $\neg$P(a) is not necessarily P(a). Differentiating between the negations $\sim$ and $\neg$, we permit not two, but three possibilities: (1) it is possible to establish that a has an attribute P; (2) it is possible to establish that a has no attribute P; (3) it is impossible to establish (accept) any of the above. The third possibility represents an empirically given fact (for example, in the case of transition states and unsolvable problems). Therefore, the term 'not-true' should be defined for the above statements in the following way: (1) P(a) is not-true if and only if $\neg$P(a) is true or

$\sim$ P(a) $\wedge$ $\sim \neg$P(a) is true; (2) $\neg$P(a) is not-true if and only if P(a) is true or if $\sim$ P(a) $\wedge$ $\sim \neg$P(a) is true.

In the case of ?P(a) the terms 'true' and 'not-true' are defined in the following manner:

(1) ?P(a) is true if and only if $\sim$ P(a) $\wedge$ $\sim \neg$P(a) is true (in other words, if P(a) is not-true and $\neg$P(a) is not-true);

(2) ?P(a) is not-true if and only if $\sim$ P(a) $\wedge$ $\sim \neg$P(a) is not-true (i.e., if P(a) is true or $\neg$ P(a) is true).

Let A, B and C be P(a), $\neg$P(a) and ?P(a) taken in arbitrary order. The relation between their truth values may be briefly expressed as follows:

(1) if A is true, then B and C are not-true;

(2) if A and B are not-true, then C is true.

Thus, two of the above cannot be both true but can be both not-true.

In the case of expressions containing quantifiers, the definitions take on the following form:

(1) ($\forall a$)x is true if and only if for any individual b such that b is a (in other words, such that $b \rightarrow a$) the expression $x(a/b)$ (where $x(a/b)$ results from the substitution of b in place of a everywhere where a occurs in x) is true;

(2) ($\neg\forall a$)x is true if and only if at least for one individual b such that b is a the expression $x(a/b)$ is not-true;

(3) (?$\forall a$)x is true if and only if $\sim$($\forall a$)x $\wedge$ $\sim$($\neg\forall a$)x is true;

(4) ($\forall a$)x is not-true if and only if ($\neg\forall a$)x $\vee$ (?$\forall a$)x is true;

(5) ($\neg\forall a$) x is not-true if and only if ($\forall a$)x $\vee$ (?$\forall a$)x is true;

(6) (?$\forall a$)x is not-true if and only if ($\forall a$)x $\vee$ ($\neg\forall a$)x is true.

For quantifier $\exists$, Definitions 7–12 are obtained in the following way: in relations 1–6 symbol $\forall$ is everywhere replaced by $\exists$; in (1) instead of the words 'for any' we use the words 'at least for one'; in (2) instead of the words 'at least for one' we use the words 'for any'.

The relation between the truth values for ($\forall a$)x, ($\neg\forall a$)x and (?$\forall a$)x is similar to that for the above expressions A, B, and C. One obtains analogous results for ($\exists a$)x, ($\neg\exists a$)x and (?$\exists a$)x.

For expressions $x \rightarrow y$ the following definitions are valid:

(1) $x \rightarrow y$ is true if and only if, having ascribed x a value 'true' (or having ascribed y a value 'not-true'), we must ascribe y the value 'true' (or x the value 'not-true');

(2) $x \to y$ is not-true if and only if the recognition of the truth of x does not lead to the recognition of the truth of y or the recognition of the non-truth of y does not lead to the recognition of the non-truth of x.

For the above statements one can introduce three and more truth values; this will affect the form of the logical theory but not its essence. In particular, this can be done in the following manner. Assume that the terms for the truth values are 'true', 'uncertain' and 'false'. Let x be a statement 'An object a may be observed'. The definitions of the terms for the truth values in the case of statements $P(a)$ and $\urcorner P(a)$ may be given the form:

(1) $P(a)$ is true if and only if $x \wedge P(a)$;
$P(a)$ is uncertain if and only if x;
$P(a)$ is false if and only if $x \wedge \urcorner P(a)$;
(2) $\urcorner P(a)$ is true if and only if $x \wedge \urcorner P(a)$;
$\urcorner P(a)$ is uncertain if and only if x;
$\urcorner P(a)$ is false if and only if $x \wedge P(a)$.

Introducing these definitions, we took into account the fact that there are three possibilities: (1) $\sim x$; (2) $x \wedge P(a)$; (3)$x \wedge \urcorner P(a)$.

But another version is also possible:

(1) $P(a)$ is true if and only if $P(a)$;
$\urcorner P(a)$ is true if and only if $\urcorner P(a)$;
$?P(a)$ is true if and only if $?P(a)$;
(2) $P(a)$ is false if and only if $\urcorner P(a)$;
$\urcorner P(a)$ is false if and only if $P(a)$;
$?P(a)$ is false if and only if $P(a) \vee \urcorner P(a)$;
(3) $P(a)$ is uncertain if and only if $?P(a)$;
$\urcorner P(a)$ is uncertain if and only if $?P(a)$.

As one can see, in this situation the uncertainty follows not from the impossibility of observing a, as in the first version, but from the impossibility of detecting P under the condition of observable a. As before, in this version there are also three (but different) possibilities: (1) $P(a)$; (2) $\urcorner P(a)$; (3) $\sim P(a) \wedge \sim \urcorner P(a)$. And if one takes into consideration both the possibility and the impossibility of observing a (as well as the possibility and the impossibility of detecting P), one obtains four possibilities. Obviously, the definitions of the truth values should be also changed, since there is an addition of one more possibility. In this case such variations as (1) introduce the fourth truth value; (2) interpret P(a) and $\neg$P(a) as uncertain not only in the case $\sim x$, but also in the case ?P(a) are possible.

Thus, truth values are not written on the statements and states themselves, to which they refer. They are introduced on the basis of taking the structure of these statements and states into account.

As is possible to demonstrate with the example of definitions for conditional statements, the definitions of truth values do not always coincide with the rules for verification of statements. Thus, in order to verify the expression $x \rightarrow y$, one has first of all to establish how has it been obtained, as through experimental research, or by means of deducing y from x, and from other statements regarded as true. In the first case the verification will consist in checking whether the rules of experimental research were observed or not, while in the second, in checking out the deduction rules.

20. Tautology, Contradictions, Realizable Statements

There are expressions which are true by virtue of the very definitions of the term 'true' (or by virtue of the rules for ascribing truth values to such expressions). These are logically true statements or tautologies. Such are, for instance, expressions of the kind $x \vee \sim x$, $\sim(x \wedge \sim x)$, $x \vee \sim x \vee y$, etc. There are also expressions which are not-true by virtue of the definitions of the term 'not-true'. These are unrealizable statements, or contradictions. Such are, for example, expressions of the type $x \vee \sim x$, $\sim(x \vee \sim x)$, $x \wedge \sim x \wedge y$, etc. Finally, there are expressions with respect to which logic is not competent to judge whether they are true or not-true. These are (what we term) 'realizable' statements. The overwhelming majority of scientific expressions are realizable.

Such an expression as $x \vee \sim x$ is true for any statements x regardless of any empirical research; it is true exclusively for the reason that it is constructed of the statements x and $\sim x$ by means of the operator $\vee$, and because there are certain rules for ascribing truth values to this kind of statement. On the other hand, such an expression as $x \wedge \sim x$ is not-true by virtue of the rules for ascribing truth values to statements with operators $\sim$ and $\wedge$. It is due to the properties of the language, and not the properties of the objective world to which x refers, that the first statement is true while the second one is not-true. The situation with other tautologies and contradictions is similar.

In these examples, the truth or non-truth of statements follows from the definitions adopted. But there are statements the truth of which cannot be established such as that of corollaries of certain definitions; nevertheless such statements are accepted in logic on the basis of different considerations in particular, as obvious. Such for instance, is the statement $a\downarrow \mathrm{P} \leftarrow \mathrm{P}$, i.e., 'An object, having a certain attribute, has this attribute'. So that the set of statements that are regarded as true by logic cannot be reduced to the set of tautologies.

Logically true statements are called the laws of logic.

First of all, one has to bear in mind that the answer to the question which statements are logically true (represent tautologies), and which are not, depends on the definitions adopted for the truth values for operators occurring in these statements (or on the definitions of these operators). Thus, it will be correct to say not simply that a statement $x \vee \sim x$ is a tautology while the statement $x \wedge \sim x$ is a contradiction, but that they appear as such under the condition of certain definitions. It is surprising how this trivial fact comes to be overlooked when it is said that some laws of two-valued logic are not preserved in many-valued, in particular, in three-valued logic. For instance, the moment one adopts definitions according to which, if x is uncertain, then $\sim x$ is uncertain, and if both x and y are uncertain, then $x \vee y$ is uncertain, one obtains the result $x \vee \sim x$ is uncertain. But it would be wrong to draw a conclusion from this to the effect that the law of two-valued logic $x \vee \sim x$ becomes non-valid in this case. The only correct conclusion here is the following: if the operators $\sim$ and $\vee$ are defined in the above fashion, and if the tautology always has a value 'true', then $x \vee \sim x$ is not a tautology.

When it is stated that the laws of two-valued logic are not preserved in many-valued logic, a logical error is made, which consists in confusion of various definitions for the terms of truth values or for logical operators.

21. Deduction

A considerable part of knowledge is acquired by people through inference (deduction) from other knowledge already available. Analysis of the rules of inference represents the most important problem of the science of logic.

Logic does not deal with arbitrary kinds of inference. Consider, for instance, statements: (1) 'A ship A has covered a distance 1000 m'; (2) 'The time spent by the ship A in this travel equals 20 hrs'. A statement (3) is inferred from the above: 'The ship A moved with an average velocity

50 m/hr'. Here statement (3) is deduced from statements (1) and (2). But this is done not according to the rules of logic, but according to a special rule for calculating the velocity (namely, the definition of the average velocity) established in physics: the value of the average velocity of a body equals the ratio of the distance covered to the time spent on it. And when in such cases it is said that a certain statement has been obtained by purely logical methods from other statements, inaccuracy is produced, or, to put it differently, confusion between deduction in the broad sense and logical deduction.

Inference (deduction) understood broadly implies obtaining of statements from certain given statements without drawing on experience (on observations and experiments); special rules are employed in this process, which are established for the linguistic symbols used in the initial statements. The set of such rules includes not only the rules of logic but other scientific rules as well, those in mathematics, physics, chemistry, etc. Inference in the more narrow sense of the word means the inference of certain statements from others exclusively by means of the rules established in the science of logic. This is logical inference, or inference according to the rules of logical entailment. For instance, from the statements 'All metals are electrically conducting' and 'Copper is a metal' one infers a statement 'Copper is electrically conducting', and this is done according to the rule of logic formulated for the quantifier 'all'. This is logical inference.

In what follows we shall drop the word 'logical' for the sake of brevity, always implying purely logical inference. The expression 'logical entailment' which a frequently used (by us as well) means the same as the expressions 'logical inference' and 'logical deduction'.

22. Logical inference

The establishing of the rules of inference represents a rather complex problem, the solution of which requires taking into consideration very different aspects of the matter. Let us outline certain common features of the inference rules.

The rules of inference are formulated in such a way that the following principle of deduction holds: if a statement y is obtained according to these rules from statements $x^1, \ldots, x^n$, and the latter are regarded as true, then the statement y should be acknowledged as true. Thus, the mysterious compulsory force of the logical laws is nothing but the force of humans themselves as applied to one of the spheres of their activity.

The rules of logical inference (entailment) represent special definitions of the properties of the logical operators occurring in the statements, as well as the consequences of such definitions. For instance, there are rules according to which from the statement $x \wedge y$ one derives logically each of x and y separately, from the statement $x \wedge y$ one derives $y \wedge x$, from $x \wedge (y \wedge z)$ one derives $(x \wedge y) \wedge z$, etc. These rules constitute on enumeration of the properties of the operator $\wedge$. The latter is introduced exactly in such a way that these rules are valid. When these rules are being formulated, the above principle of deduction is taken into account. Indeed, if $x \wedge y$ is true, then x is true, etc.

The logical operators are introduced not on an individual basis but as a set, since in statements they also occur as a set. For instance, an expression $(\forall a)(x \vee (\exists b) \sim (y \wedge z))$ contains operators $\forall$, $\exists$, $\vee$, $\sim$, $\wedge$. And, certainly, the rules of inference for such expressions should account for all possible combinations of these operators. Thus, although each operator considered as an individual appears trivially simple, the definition of their properties in various combinations and in consideration of all possible combinations of this kind is a quite difficult matter.

The answer to the question, when from some statements other statements logically follow, is provided not by general reasoning but by enumeration of concrete cases. Therefore, a convenient method for this was represented by the logical calculi which not only provided a rigorous formulation of the above rules but frequently allowed their exhaustive description.

The fact that from a statement x one can logically deduce a statement y (a statement y logically follows from a statement x) will be denoted by a symbol

$$x \vdash y.$$

In this expression the symbol $\vdash$ is not a logical operator. It is the predicate of a statement 'From a statement x logically follows a statement y' (the predicate of entailment). The subjects of this statement are the terms 'a statement x' and 'a statement y'. The statement x is called the premise, the statement y, the conclusion, or the consequence.

If one ignores the structure of the premises and the conclusions, the predicate of entailment possesses only the following properties:

(1) if $x \vdash y$, and x is accepted (recognized, regarded) as true, then y is accepted (regarded, must be recognized) as true;

(2) if $x \vdash y$ and y is rejected (regarded as not-true), then x is rejected;

(3) if $x \vdash y$ and $y \vdash z$, then $x \vdash z$.

All other properties of the predicate of entailment arise because of the properties of terms, statements and operators occurring in the premises and the conclusions. Those are not its properties proper but rather the properties of premises and conclusions. The predicate of entailment is used only as a means for establishing these properties.

If inference is made from several statements $x^1, \ldots, x^n$, then we may agree on the following: from $x^1, \ldots, x^n$ logically follows y if and only if from $x^1 \wedge \ldots \wedge x^n$ logically follows y.

Such symbols as

$$x \dashv\vdash y$$

will denote that both $x \vdash y$ and $y \vdash x$ hold. In this case we shall consider x and y as deductively equivalent.

Such symbols as

$$\vdash x$$

will denote the fact that x is accepted from purely logical considerations (as logically true). Here $\vdash$ is the predicate of the statement 'A statement x is logically true'. We are using the same symbol as that for entailment, since each time it would be clear from the context which is implied. Obviously, if $\vdash x$, then x is true.

If one neglects the structure of x, then the predicate $\vdash$ in the expressions of the form $\vdash x$ (in this case it will called the predicate of logical truth) has no other properties, except the one given above. In combination with the predicate of entailment, the predicate of logical truth is characterized by the following properties:

(1) if $x \vdash y$ and $\vdash x$, then $\vdash y$;
(2) if $x \vdash y$ and y is not logically true, then x is not logically true.

The other properties of the predicate of logical truth are due to the fact that it becomes a means for defining the properties of logical operators as well as of the terms and statements containing them.

The predicate of logical truth may be regarded as the predicate of entailment with an empty premise, i.e., as the predicate of degenerate entailment.

23. The General Theory of Deduction

The alphabet of the general theory of deduction (GTD):

(1) propositional variables or variables for statements;

(2) variables for subject-terms;
(3) variables for predicate-terms;
(4) $\wedge$, $\vee$, $\sim$, $\urcorner$, ?, $\rightarrow$, $\leftrightarrow$ – the operators of conjunction, disjunction, external and internal negations, uncertainty, conditionality and conditional equivalence, respectively;
(5) $\vdash$ – the predicates of logical entailment and logical truth.

D1. Definition of a propositional formula (or a statement form):

(1) the propositional variables are propositional formulas;
(2) if x and y are propositional formulas, then $(x \wedge y)$, $(x \vee y)$, $(x \rightarrow y)$ and $(x \leftrightarrow y)$ are propositional formulas; if $x^1, x^2, \ldots, x^n$ $(n \geqslant 3)$ are propositional formulas, then $(x^1 \wedge x^2 \wedge \ldots \wedge x^n)$ and $(x^1 \vee x^2 \vee \ldots \vee x^n)$ are propositional formulas;
(3) if a is a variable for subjects, and P is a variable for predicates, then $P(a)$, $\urcorner P(a)$ and $?P(a)$ are propositional formulas;
(4) if a is a variable for terms-subjects, and x is a variable for statements, then $(\forall a)x$, $(\urcorner\forall a)x$, $(?\forall a)x$, $(\exists a)x$, $(\urcorner\exists a)x$ and $(?\exists a)x$ are propositional formulas;
(5) if x is a propositional formula, then $\sim x$ is a propositional formula;
(6) something is a propositional formula of GTD only by virtue of (1)–(5).

D2. Definition of the formula of logical entailment: $x \vdash y$ is a formula of logical entailment if and only if x and y are propositional formulas.

D3. Definition of the formula of logical truth (or degenerate entailment): $\vdash x$ is a formula of logical truth if and only if x is a propositional formula.

Such expressions as $x \dashv\vdash y$ will be used for abbreviation of two formulas $x \vdash y$ and $y \vdash x$. The brackets will frequently be omitted, under an assumption that $\wedge$ binds more strongly than $\vee$; both of them bind more strongly than $\rightarrow$, $\leftrightarrow$ and $\vdash$; $\rightarrow$ and $\leftrightarrow$ bind more strongly than $\vdash$ and $\dashv\vdash$.

Axiomatic schemes of GTD:

(I)

(1) $\sim\sim x \dashv\vdash x$;
(2) $x^1 \wedge x^2 \wedge \ldots \wedge x^n \vdash x^i$,

where x^i is any of $x^1, x^2, \ldots, x^n$, and $n \geqslant 2$;

(3) $x^1 \wedge x^2 \wedge \ldots \wedge x^n \dashv\vdash y$,

where y differs from $x^1 \wedge x^2 \wedge \ldots \wedge x^n$ only in that its certain part of the form $x^{il} \wedge \ldots \wedge x^{ik}$ is included in brackets;

(4) $x \wedge y \vdash y \wedge x$;

(5) $(x^1 \vee x^2 \vee \ldots \vee x^n) \wedge y \dashv\vdash (x^1 \wedge y) \vee (x^2 \wedge y) \vee \ldots \vee (x^n \wedge y)$;

(6) $\sim(x^1 \wedge x^2 \wedge \ldots \wedge x^n) \dashv\vdash \sim x^1 \vee \sim x^2 \vee \ldots \vee \sim x^n$;

(7) $(x \vee y) \wedge \sim x \vdash y$;

(8) $\vdash (x \vee \sim x)$

(II)

(1) $(x \rightarrow y) \wedge x \vdash y$;

(2) $(x \rightarrow y) \vdash (\sim y \rightarrow \sim x)$;

(3) $(x \rightarrow y) \wedge (y \rightarrow z) \vdash (x \rightarrow z)$;

(4) $(x \rightarrow y \wedge z) \dashv\vdash (x \rightarrow y) \wedge (x \rightarrow z)$;

(5) $(x \leftrightarrow y) \dashv\vdash (x \rightarrow y) \wedge (y \rightarrow x)$;

(III)

(1) $\mathrm{P}(a) \dashv\vdash \sim \neg\mathrm{P}(a) \wedge \sim ?\mathrm{P}(a)$;

(2) $\neg\mathrm{P}(a) \dashv\vdash \sim \mathrm{P}(a) \wedge \sim ?\mathrm{P}(a)$;

(3) $?\mathrm{P}(a) \dashv\vdash \sim \mathrm{P}(a) \wedge \sim \neg\mathrm{P}(a)$;

(IV)

(1) $(\forall a)x \vdash x$;

(2) $x \vdash (\exists a)x$;

(3) $(\forall a)x \dashv\vdash (\neg\exists a) \sim x$;

(4) $(\neg\forall a)x \dashv\vdash (\exists a) \sim x$;

(5) $(?\forall a)x \dashv\vdash (?\exists a) \sim x$;

(6) $(\forall a)x \dashv\vdash \sim(\neg\forall a)x \wedge \sim(?\forall a)x$;

(7) $(\neg\forall a)x \dashv\vdash \sim(\forall a)x \wedge \sim(?\forall a)x$;

(8) $(?\forall a)x \dashv\vdash \sim(\forall a)x \wedge \sim(\neg\forall a)x$;

(9) $(\forall a)(x \wedge y) \dashv\vdash (\forall a)x \wedge (\forall a)y$;

(10) $(\exists a)(x \vee y) \dashv\vdash (\exists a)x \vee (\exists a)y$;

(11) $(\forall a)(x \vee y) \vdash (\forall a)x \vee (\exists a)y$;

(12) $(\exists a)(x \wedge y) \vdash (\exists a)x \wedge (\exists a)y$;

(13) $(\forall a)x \vee (\forall a)y \vdash (\forall a)(x \vee y)$;

(14) $(\forall a)x \wedge (\exists a)y \vdash (\exists a)(x \wedge y)$;

The rules of inference of GTD:

(1) If $x \vdash y$ and $y \vdash z$, then $x \vdash z$;

(2) If $x \vdash y$ and $x \vdash z$, then $x \vdash y \wedge z$;
(3) If $w \wedge x \vdash y$, $w \wedge y \vdash x$ and v is formed of z through substitution of y in place of one or more occurrences of x in z as a statement, then $w \wedge z \vdash v$;
(4) If $x \vdash y$ and $\vdash x$, then $\vdash y$;
(5) If $x \vdash y$ and $\vdash \sim y$, then $\vdash \sim x$;
(6) If $\vdash x$ and $\vdash y$, then $\vdash x \wedge y$;
(7) If $x \vdash y$, then $(\forall a)x \vdash (\forall a)y$;
(8) If $x \vdash y$, then $(\exists a)x \vdash (\exists a)y$;
(9) If $\vdash x$, then $\vdash (\forall a)x$;
(10) If $x \vdash y$, then $\vdash (x \rightarrow y)$.

D4. Definition of a provable formula of GTD: a formula of logical entailment or logical truth is provable in GTD (represents a theorem) if and only if it is either an axiom or is obtained from provable formulae (from theorems) according to the rules of inference of GTD.

The substitution of statements and terms, respectively, in place of variables in propositional formulae results in statements, while this kind of substitution in the formulae of logical entailment and logical truth results in assertions to the effect that from some statements logically follow others or that the statements are logically true.

24. Classical and Non-Classical Cases in the Theory of Inference

In the above theory of inference two forms of negation $\sim$ and $\urcorner$ are distinguished, and the uncertainty operator appears. The following assertions:

(1) $\sim P(a) \vdash \urcorner P(a)$, $\sim (\forall a)x \vdash (\urcorner \forall a)x$;
$\sim (\exists a)x \vdash (\urcorner \exists a)x$;
(2) $\vdash P(a) \vee \urcorner P(a)$, $\vdash (\forall a)x \vee (\urcorner \forall a)x$;
$\vdash (\exists a)x \vee (\urcorner \exists a)x$;

are unprovable in it.

This case will be called the non-classical, or the general case of the theory of inference.

But there are also instances, when the two forms of negation are not discriminated, and the uncertainty operator appears superfluous. In such cases the theory of inference may be expanded on the basis of the addition of rule 1. Then assertions 2 will become provable. Therefore, the rules of predication as well as the rules 5–8 for quantifiers disappear as redundant. Rules 3 and 4 for quantifiers will be equivalent to the following ones:

$$(\forall a)x \dashv\vdash \sim(\exists a)\sim x.$$

In the theory of inference, we shall call this case the classical one.

Let us mention still one more most important difference between the classical and non-classical cases. In the classical case the conjunction of negations of both terms in the pairs $\mathrm{P}(a)$ and $\neg\mathrm{P}(a)$, $(\forall a)x$ and $(\neg\forall a)x$, $(\exists a)x$ and $(\neg\exists a)x$ results in contradiction. We shall demonstrate this with the example of the first pair: since

$$\neg\mathrm{P}(a) \dashv\vdash \sim\mathrm{P}(a),\ \sim\neg\mathrm{P}(a) \dashv\vdash \sim\sim\mathrm{P}(a),\ \sim\sim\mathrm{P}(a) \dashv\vdash \mathrm{P}(a),$$

we obtain

$$\sim\mathrm{P}(a) \wedge \sim\neg\mathrm{P}(a) \vdash \sim\mathrm{P}(a) \wedge \sim\sim\mathrm{P}(a),$$
$$\sim\mathrm{P}(a) \wedge \sim\neg\mathrm{P}(a) \vdash \sim\mathrm{P}(a) \wedge \mathrm{P}(a).$$

On the other hand, in the non-classical case we do not end up in a contradiction since the substitution of $\sim\sim\mathrm{P}(a)$ in place of $\sim\neg\mathrm{P}(a)$ with the subsequent substitution of $\mathrm{P}(a)$ is inadmissible here. The situation for other pairs is similar.

The neglect of this difference leads to confusion. This differentiation represents the differentiation between the cases when two possibilities and three possibilities are permitted. In the non-classical case these possibilities are: $\mathrm{P}(a)$, $\neg\mathrm{P}(a)$ and $?\mathrm{P}(a)$, in the classical, only $\mathrm{P}(a)$ and $\neg\mathrm{P}(a)$. The situation with quantifiers is similar. The non-classical case is more general, since, as a special case, one may assume here that the third possibility with the uncertainty operator is empty.

25. The Rules of Inference and the Truth Values of Statements

The rules of inference are independent of the number of truth values one decides to ascribe to statements. When inference is realized, we take into consideration just the apparent structure of statements, i.e., the presence in them of certain terms, statements and logical operators as well as their mutual arrangement. On the other hand, truth values are taken into account only for the realization of the main principle of deduction: if the premises are accepted (regarded as true), then the conclusions are accepted; if the conclusions are rejected (regarded as not-true), then the premises are rejected. In this case it would be quite sufficient to introduce one truth predicate together with its negation.

26. Identity by Meaning and Entailment

Let us adopt the rule: if $x \equiv y$, then $x \vdash y$ and $y \vdash x$. Thus, if one adopts the definition $x \equiv$ Df. y, one accepts $x \equiv y$, which means that two assertions $x \vdash y$ and $y \vdash x$ are accepted. Because of this, for the sake of brevity, we can accept as definitions the statements concerning entailment.

27. The General Theory of Terms

The general theory of terms (GTT) establishes the properties for predicates $\rightharpoondown$, $\rightleftharpoons$ and $\equiv$, and also the properties of T-operators and the terms containing them, without taking into account the concrete meaning of terms. Below we shall give a system of GTT which represents a superstructure over GTD (i.e., the latter is assumed).

The alphabet of GTT:

(1) object variables;
(2) variables for terms;
(3) subject variables;
(4) predicate variables;
(5) S – the predicate of denotation;
(6) $\rightharpoondown$ – the predicate of inclusion by meaning;
(7) $\rightleftharpoons$ – the predicate of identity by meaning;
(8) $\downarrow$ – the operator 'which', 'such, that', 'the fact, that';
(9) (...) – the operator on *n*-tuple of terms (pair, triplet,...)
(10) m – the quasioperator of metaterm;
(11) s – the 'subject' term;
(12) P – the 'attribute' term.

D1. Definition of the subject form (SF):

(1) the subject variables are SF;
(2) if $a^1, \ldots, a^n$ are SF, then $(a^1, \ldots, a^n)$ is an SF;
(3) if $a^1, \ldots, a^n$ are SF, then $(a^1 \vee \ldots \vee a^n)$, $(a^1 \wedge \ldots \wedge a^n)$, $(\vee\ a^1, \ldots, a^n)$ and $(\wedge\ a^1, \ldots, a^n)$ are SF;

(4) if a is SF, then $\sim a$ is SF;
(5) if a is SF, and b is a predicate form, then $a{\downarrow}\alpha b$ and $a{\downarrow}\sim\alpha b$ are SF (here α denotes the presence or absence of $\urcorner$ or ?);
(6) if b is a predicate form, the ${\downarrow}\alpha b$ and ${\downarrow}\sim\alpha b$ are SF;
(7) if a is SF, and x is a propositional formula, then $a{\downarrow}x$ is SF;
(8) if x is a propositional formula, then ${\downarrow}x$ is SF;
(9) if a is either SF or a predicate form or a propositional formula, them ma is SF;
(10) s is SF;
(11) if a is SF, then ia is SF ($i = 1, 2, 3, \ldots$);
(12) something is SF only by virtue of (1)–(11).

D2. Definition of the predicate form (PF):

(1) the predicate variables are PF;
(2) if $a^1, \ldots, a^n$ are PF, then $(a^1, \ldots, a^n)$, $(a^1 \wedge \ldots \wedge a^n)$, $(a^1 \vee \cdots \vee a^n)$, $(\wedge a^1, \ldots, a^n)$, $(\vee a^1, \ldots, a^n)$ are PF;
(3) if a is PF, then $\sim a$ is PF;
(4) if a is SF, and b is PF, then $b{\downarrow}\alpha a$ and $b{\downarrow}\alpha a$ are PF;
(5) if a is SF, then ${\downarrow}\alpha a$ and ${\downarrow}\sim\alpha e$ are PF;
(6) if a is PF, and x is a propositional formula, then $a{\downarrow}x$ is PF;
(7) if x is a propositional formula, then $x{\downarrow}$ is PF;
(8) S, $\rightharpoonup$, $\rightleftharpoons$ are PF;
(9) P is PF;
(10) something is PF only by virtue of (1)–(9).

D3. Definition of the object variable: ia is an object variable if and only if a is a subject variable ($i = 1, 2, 3, \ldots$).

D4. SF and PF, and they only, are term forms (TF).

D5. Addition to the definition of the propositional formula (FP):

(1) if a is SF, and P is PF, then αP(a) is FP (α denotes the presence or the absence of $\urcorner$ or ?);
(2) S(a, mb) is FP if and only if a and b are both SF;
(3) $\rightharpoonup$(ma, mb) and $\rightleftharpoons$ (ma, mb) are FP if and only if both a and b are SF (simultaneously) or PF (simultaneously).

In order to simplify the notation and to obtain a greater visual clarity, we shall use, instead of the symbols $\rightharpoonup$(m*a*, m*b*) and $\rightleftharpoons$(m*a*, m*b*), notations $a \rightharpoonup b$ and $a \rightleftharpoons b$, respectively.

D6. Expression *a*, which is either TF or FP, occurs in expression *b*, which is either TF or FP, as TF or FP if *a* is a part of *b*, with the following exceptions:

(1) *a* does not represent an occurrence in m*a* as TF or FP;
(2) if *c* is a part of *a*, then *c* does not represent an occurrence in m*a* as TF or FP.

Let us explain the meaning of symbols that are specially used in the theory of terms:

(1) S(*a*, m*b*) stands for '*a* is denoted by the term *b*';
(2) $\rightharpoonup$(m*a*, m*b*) – 'The term *a* is subordinated by value to the term *b*' (or 'Everything, denoted by the term *a*, is denoted also by the term *b*');
(3) $\rightleftharpoons$(m*a*, m*b*) – 'The terms *a* and *b* are identical by value';
(4) $(a^1, \ldots, a^n)$ – 'A pair (triplet, quadruplet, etc., in general – *n*-tuple) of objects';
(5) $(a \wedge b)$ – 'An object (attribute), denoted by the term *a* and the term *b*';
(6) $(a \vee b)$ – 'An object (attribute), denoted by at least one of the terms *a* and *b*';
(7) $\sim a$ – 'Not-*a*' in the sense 'An object, which is not denoted by the term *a*';
(8) $(\wedge\, ab)$ – 'Each of *a* and *b*';
(9) $(\vee\, ab)$ – 'At least one of *a* and *b*';
(10) $(\wedge\, ab)$ – 'Not each of *a* and *b*';
(11) $(\vee\, ab)$ – 'Neither of *a* and *b*';
(12) $\tilde{a}$ – 'Not-*a*' in the sense 'having no attribute *a*';
(13) $a{\downarrow}x$ – '*a* such that *x*', '*a*, which is *x*', '*a*, having *x*' etc.;
(14) ${\downarrow}x$ – 'The fact that *x*', 'having *x*', etc.;
(15) $x{\downarrow}$ – 'Such that *x*', 'is characterized by the fact that *x*', etc.

The object variables are the variables for subject-terms, under the condition that the latter possess the following properties: each of *ia* separately reads as 'any *a*' ('a certain, but it is indifferent, which one, *a*', etc.); if *ia* and *ka* occur together, they are read as 'any other *a*'.

Axiomatic schemes of GTT:

(I)

(1) $\vdash S(a, \mathrm{m}a)$;
(2) $\sim S(a, \mathrm{m}b) \vdash \neg S(a, \mathrm{m}b)$.

(II)

(1) $(a \rightharpoondown b) \dashv\vdash (S(s, \mathrm{m}a) \rightarrow S(s, \mathrm{m}b))$;

where a and b are SF;

(2) $(a \rightharpoondown b) \dashv\vdash (s \downarrow a \rightharpoondown s \downarrow b)$.

where a and b are PF;

(3) $(a \rightleftharpoons b) \dashv\vdash (a \rightharpoondown b) \wedge (b \rightharpoondown a)$

(III)

(1) $\vdash a \rightleftharpoons s \downarrow S(s, \mathrm{m}a)$;
(2) $S(s, \mathrm{m} \sim a) \dashv\vdash \sim S(s, \mathrm{m}a)$;
(3) $S(s, \mathrm{m}(a^1 \wedge \ldots \wedge a^n)) \dashv\vdash S(s, \mathrm{m}a^1) \wedge \ldots \wedge S(s, \mathrm{m}a^n)$;
(4) $S(s, \mathrm{m}(a^1 \vee \ldots \vee a^n)) \dashv\vdash S(s, \mathrm{m}a^1) \vee \ldots \vee S(s, \mathrm{m}a^n)$;

(IV)

(1) $\alpha P(\wedge\, a^1 \ldots a^n) \dashv\vdash \alpha P(a^1) \wedge \ldots \wedge \alpha P(a^n)$;
(2) $\alpha(\wedge\, P^1 \ldots P^n)(a) \dashv\vdash \alpha P^1(a) \wedge \ldots \wedge \alpha P^n(a)$;
(3) $\alpha P(\vee\, a^1 \ldots a^n) \dashv\vdash \alpha P(a^1) \vee \ldots \vee \alpha P(a^n)$;
(4) $\alpha(\vee\, P^1 \ldots P^n)(a) \dashv\vdash \alpha P^1(a) \vee \ldots \vee \alpha P^n(a)$;
(5) $\alpha P(\tilde{\wedge}\, a^1 \ldots a^n) \dashv\vdash \sim \alpha P(\wedge\, a^1 \ldots a^n)$;
(6) $\alpha P(\tilde{\vee}\, a^1 \ldots a^n) \dashv\vdash \sim \alpha P(\vee\, a^1 \ldots a^n)$;
(7) $\alpha(\tilde{\wedge}\, P^1 \ldots P^n)(a) \dashv\vdash \sim \alpha(\wedge\, P^1 \ldots P^n)(a)$;
(8) $\alpha(\vee\, P^1 \ldots P^n)(a) \dashv\vdash \sim \alpha(\vee\, P^1 \ldots P^n)(a)$;
(9) $\tilde{P}(a) \dashv\vdash \sim P(a)$;

(V)

(1) $(a^1, \ldots, a^n) \rightharpoondown (b^1, \ldots, b^n) \dashv\vdash (a^1 \rightharpoondown b^1) \wedge \ldots \wedge (a^n \rightharpoondown b^n)$;
(2) $(\forall (a^1, \ldots, a^n))x \dashv\vdash (\forall a^1)x \wedge \ldots \wedge (\forall a^n)x$;
(3) $(\exists (a^1, \ldots, a^n))x \dashv\vdash (\exists a^1)x \wedge \ldots \wedge (\exists a^n)x$;

(VI)

(1) $\vdash a \downarrow (x \wedge y) \rightleftharpoons (a \downarrow x) \wedge (a \downarrow y)$;

(2) $\vdash a{\downarrow}(x \vee y) \rightleftharpoons (a{\downarrow}x) \vee (a{\downarrow}y)$;
(3) $\vdash (s{\downarrow} \sim x) \rightleftharpoons \sim (s{\downarrow}x)$;
(4) $({\downarrow}x \rightleftharpoons {\downarrow}y) \dashv\vdash (x{\downarrow} \rightleftharpoons y{\downarrow})$;
(5) $\vdash (a{\downarrow}x){\downarrow}y \rightleftharpoons (a{\downarrow}y){\downarrow}x)$;
(6) $\vdash x \rightarrow (a{\downarrow}x \rightharpoonup a)$;
(7) $(\forall{\downarrow}x)(y{\downarrow})(x{\downarrow}) \dashv\vdash (x \rightarrow y)$;
(8) $(\neg\forall{\downarrow}x(y{\downarrow})({\downarrow}x) \dashv\vdash (x\neg\rightarrow y)$;

(VII)

(1) $\vdash (\exists a)x \rightarrow (x{\downarrow})(a{\downarrow}x)$;
(2) $\vdash (\exists a)\alpha P(a)\alpha P(a{\downarrow}\alpha P(a))$;
(3) $\vdash (\exists a) \sim \alpha P(a) \rightarrow \sim \alpha P(a{\downarrow} \sim \alpha P(a))$,

where α denotes the presence or absence of $\neg$ or ?.

(VIII)

(1) $(\exists a)x \wedge ((b \rightharpoonup a) \rightarrow (a \rightharpoonup b)) \vdash x$;
(2) $x \wedge ((b \rightharpoonup a) \rightarrow (a \rightharpoonup b)) \vdash (\forall a)x$;
(3) $\sim (a \rightharpoonup b^1) \wedge \ldots \wedge \sim (a \rightharpoonup b^n) \wedge (\exists a)x \vdash (\forall a)x$,

where a does not occur as free in x as SF, and $b^1, \ldots, b^n$ are all the SF with free occurrences in x as SF.

(IX)

(1) $(a \rightharpoonup b) \wedge (\forall b)x \vdash (\forall a)y$,

where y is formed of x through substitution of a in place of all free occurrences of b in x as SF.

(2) $(a \rightharpoonup b) \wedge (\exists a)x \vdash (\exists b)y$,

where y is formed of x through substitution of b in place of all free occurrences of a in x as SF.

(3) $(a \rightleftharpoons b) \wedge x \vdash y$,

where y is formed of x through substitution of b in place of one or more occurrences of a in x as TF.

(X)

(1) $(\forall a)x \wedge (a \rightharpoonup b) \vdash (\forall a)y$,

where y is formed of x through substitution of a in place of b everywhere,

where b occurs as free in x as SF; in the subsequent schemes (2)–(4) the relation between x and y is similar;

(2) $(\forall a)y \wedge (a \rightharpoonup b) \vdash (\forall a)x$;
(3) $(\exists a)x \wedge (a \rightharpoonup b) \vdash (\exists a)y$;
(4) $(\exists a)y \wedge (a \rightharpoonup b) \vdash (\exists a)x$;

(XI)

(1) $(\forall a)x \wedge (b \rightharpoonup a) \vdash (\forall b)x$,

where a does not occur as free in x as SF; in the subsequent schemes (2)–(4) the relation between a and x is similar;

(2) $(\forall b)x \wedge (b \rightharpoonup a) \vdash (\forall a)x$;
(3) $(\exists a)x \wedge (b \rightharpoonup a) \vdash (\exists b)x$;
(4) $(\exists b)x \wedge (b \rightharpoonup a) \vdash (\exists a)x$;

(XII)

(1) $(\mathrm{P} \rightharpoonup \mathrm{Q}) \wedge \mathrm{P}(a) \vdash \mathrm{Q}(a)$;
(2) $(\mathrm{P} \rightharpoonup \mathrm{Q}) \wedge \urcorner \mathrm{Q}(a) \vdash \urcorner \mathrm{P}(a)$;
(3) $(\mathrm{P} \rightharpoonup \mathrm{Q}) \wedge \sim \mathrm{Q}(a) \vdash \sim \mathrm{P}(a)$;

(XIII)

(1) $\alpha \mathrm{P}(a) \dashv\vdash \mathrm{S}(a, \mathrm{m}(s{\downarrow}\, \alpha \mathrm{P}))$;
(2) $\sim \alpha \mathrm{P}(a) \dashv\vdash \mathrm{S}(a, \mathrm{m}(s{\downarrow} \sim \alpha \mathrm{P}))$;
(3) $(\forall a)\alpha \mathrm{P}(a) \dashv\vdash (a \rightharpoonup s{\downarrow}\, \alpha \mathrm{P})$;
(4) $(\forall a) \sim \alpha \mathrm{P}(a) \dashv\vdash (a \rightharpoonup s{\downarrow} \sim \alpha \mathrm{P})$;
(5) $(\exists a)\alpha \mathrm{P}(a) \dashv\vdash (\exists(s{\downarrow}\, \alpha \mathrm{P}))(\forall p{\downarrow}\, a)(p{\downarrow}\, a)(s{\downarrow}\, \alpha \mathrm{P})$;
(6) $(\exists a) \sim \alpha \mathrm{P}(a) \dashv\vdash (\exists(s{\downarrow} \sim \alpha \mathrm{P}))(\forall p{\downarrow}\, a)(p{\downarrow}\, a)(s{\downarrow} \sim \alpha \mathrm{P})$;
(7) $(\forall a)\alpha \mathrm{P}(a) \dashv\vdash (\forall(s{\downarrow} \sim \alpha \mathrm{P}))(\exists p{\downarrow}\, a)\urcorner(p{\downarrow}\, a)(s{\downarrow} \sim \alpha \mathrm{P})$;
(8) $(\forall a) \sim \alpha \mathrm{P}(a) \dashv\vdash (\forall(s{\downarrow}\, \alpha \mathrm{P}))(\exists p{\downarrow}\, a)\urcorner(p{\downarrow}\, a)(s{\downarrow}\, \alpha \mathrm{P})$;

(XIV)

(1) $\vdash ({\downarrow}\, x \rightleftharpoons s{\downarrow}\, x)$;
(2) $\vdash (x{\downarrow} \rightleftharpoons p{\downarrow}\, x)$;
(3) $\vdash (\tilde{x}{\downarrow} \rightleftharpoons (\sim x){\downarrow})$;
(4) $\vdash a{\downarrow}\, \alpha \mathrm{P} \rightleftharpoons a{\downarrow}\, \alpha \mathrm{P}(a)$;
(5) $\vdash a{\downarrow} \sim \alpha \mathrm{P} \rightleftharpoons a{\downarrow} \sim \alpha \mathrm{P}(a)$;
(6) $\vdash \mathrm{P}{\downarrow}\, \alpha a \rightleftharpoons \mathrm{P}{\downarrow}\, \alpha \mathrm{P}(a)$;
(7) $\vdash \mathrm{P}{\downarrow} \sim \alpha a \rightleftharpoons \mathrm{P}{\downarrow} \sim \alpha \mathrm{P}(a)$;

(8) $\vdash \downarrow \alpha P \rightleftharpoons s \downarrow \alpha P$;
(9) $\vdash \downarrow \sim \alpha P \rightleftharpoons s \downarrow \alpha P$;
(10) $\vdash \downarrow \alpha a \rightleftharpoons p \downarrow \alpha a$;
(11) $\vdash \downarrow \sim \alpha a \rightleftharpoons p \downarrow \sim \alpha a$;
(12) $(\forall a)\alpha P(a \downarrow x) \dashv\vdash (\forall a \downarrow x)\alpha P(a \downarrow x)$;
(13) $(\exists a)\alpha P(a \downarrow x) \dashv\vdash (\exists a \downarrow x)\alpha P(a \downarrow x)$.

(XV)

(1) $\vdash S(a, \mathrm{m}s)$,

where a is SF, and s does not occur in a;

(2) $\vdash S(s \downarrow a), \mathrm{m}(s \downarrow p))$,

where a is PF, and p does not occur in a;

(3) $\vdash ia \rightharpoonup a$;
(4) $(ia \rightleftharpoons ib) \dashv\vdash (a \rightleftharpoons b)$.

The rules of GTT:

(1) if $x \dashv\vdash y$, then $\vdash a \rightleftharpoons b$, where b is formed of a through substitution of y in place of one or more occurrences of x in a as FP.
(2) If $\vdash (a \rightleftharpoons b)$, then $x \dashv\vdash y$, where y is formed of x through substitution of b in place of one or more occurrences of a in x.
(3) If $\vdash \downarrow x \rightleftharpoons \downarrow y$, then $x \dashv\vdash y$.
(4) If $a \rightharpoonup b$ and $\vdash x$, then $\vdash (\exists ib)y$, where y is formed of x through substitution of ib in place of one or more free occurrence of a in x as SF, and ib does not occur in x.
(5) If $a \rightharpoonup b$ and $\vdash x$, then $\vdash (\forall ib)y$, where y is formed of x through substitution of ib in place of all free occurrences of a in x as SF, and ib does not occur in x.

Let us list a few theorem schemes of GTT:

(1) $\vdash (a \rightharpoonup a)$;
(2) $(a \rightharpoonup b) \wedge (b \rightharpoonup c) \vdash (a \rightharpoonup c)$;
(3) $\vdash \sim a \rightleftharpoons s \downarrow \sim S(s, \mathrm{m}a)$,

where a is SF.

(4) $\vdash \sim a \rightleftharpoons p \downarrow \sim S((s \downarrow p), \mathrm{m}a)$,

where a is PF;

(5) $\vdash (a \wedge b) \rightleftharpoons s{\downarrow}(\mathrm{S}(s, \mathrm{m}a) \wedge \mathrm{S}(s, \mathrm{m}b))$,

where a and b are SF;

(6) $\vdash (a \wedge b) \rightleftharpoons p{\downarrow}(\mathrm{S}((s{\downarrow} p), \mathrm{m}(s{\downarrow} a)) \wedge \mathrm{S}((s{\downarrow} p), \mathrm{m}(s{\downarrow} b)))$,

where a and b are PF;

(7) $\vdash (a \vee b) \rightleftharpoons s{\downarrow}(\mathrm{S}(s, \mathrm{m}a) \vee \mathrm{S}(s, \mathrm{m}b))$,

where a and b are SF;

(8) $\vdash (a \vee b) \rightleftharpoons p{\downarrow}(\mathrm{S}((s{\downarrow} p), \mathrm{m}(s{\downarrow} a)) \vee \mathrm{S}((s{\downarrow} p), \mathrm{m}(s{\downarrow} b)))$,

where a and b are PF;

(9) $(a \rightharpoonup b) \vdash (\forall a)(\forall b)(a \rightharpoonup b)$;
(10) $\vdash \sim\sim a \rightleftharpoons a$;
(11) $\vdash (a \wedge b) \rightharpoonup a$;
(12) $\vdash (a \wedge b) \rightharpoonup (b \wedge a)$;
(13) $\vdash (a \wedge b \wedge c) \rightleftharpoons ((a \wedge b) \wedge c)$;
(14) $\vdash (a \vee b) \rightleftharpoons \sim(\sim a \wedge \sim b)$;
(15) $\vdash \sim(a \wedge b) \rightleftharpoons (\sim a \vee \sim b)$;
(16) $\vdash (a \vee a) \rightleftharpoons a$;
(17) $\vdash (a \wedge a) \rightleftharpoons a$;
(18) $\vdash a \rightharpoonup (a \vee b)$;
(19) $(a \rightharpoonup b) \vdash (\sim b \rightharpoonup a)$;
(20) $(a \rightharpoonup b) \wedge (a \rightharpoonup c) \vdash (a \rightharpoonup b \wedge c)$;
(21) $(a \rightharpoonup b) \vee (a \rightharpoonup c) \vdash (a \rightharpoonup b \vee c)$;
(22) $(\forall a)x \wedge (\forall a{\downarrow} x)(y{\downarrow})({\downarrow} x) \vdash (\forall a)y$;
(23) $(\exists a)x \wedge (\forall a{\downarrow} x)(y{\downarrow})({\downarrow} x) \vdash (\exists a)y$;
(24) $x \wedge (\forall a{\downarrow} x)(y{\downarrow})({\downarrow} x) \vdash y$;
(25) $\vdash (a{\downarrow} \sim \mathrm{P}) \rightleftharpoons (a{\downarrow} \neg \mathrm{P}) \vee (a{\downarrow} ?\mathrm{P})$;
(26) $\vdash a{\downarrow} \sim\sim \mathrm{P} \rightleftharpoons a{\downarrow} \mathrm{P}$;
(27) $\vdash \mathrm{P}(a{\downarrow} \mathrm{P})$;
(28) $\vdash (a \rightleftharpoons b) \rightarrow (c \rightleftharpoons d)$,

where d is formed of c through substitution of b in place of one or more occurrences of a in c.

Certainly, in what follows we shall not need the above system of GTT in its entirety. But we believed it appropriate to present it here, our purpose being to demonstrate the general orientation of the theory of terms.

28. The Coordinates of Statements

Let us adopt the following additions to the general theory of inference and terms, which will make them explicitly applicable to the languages of experimental sciences.

If x is a statement (or a term), and α is a linguistic expression denoting a spatial or temporal domain (or both), then such expression as 'x in α' will be also considered as a statement (respectively, a term). The expression α will be called the coordinates of x. In abbreviated form, the statements and terms of this kind will be denoted by symbols $[x\alpha]$. No logical relation between x and α in $[x\alpha]$ is implied (although, as a special case, it is not excluded). We shall say that in $[x\alpha]$ the statement (or term) x is bound by coordinates α.

If x is a statement or a term, and y is a statement, then 'x for y' (or 'x under the condition that y') represents a statement, or, respectively, a term. We shall call y the condition for x. For the sake of brevity, such statements and terms will be written in the form $[xy]$. We shall say that x in $[xy]$ is bound by condition y.

No logical relation between x and y in $[xy]$ is implied in the general case, in other words, from $[xy]$, where x is a statement, the expressions $x \to y$ and $y \to x$ do not follow logically. On the other hand, if x is a term, the absence of logical relation between x and y is obvious (since y is a statement).

The theory of logical coordinates (TLC) is constructed as an addition to GTD and GTT.

The alphabet of TLC:

(1) variables for the terms of spatial coordinates $s, s^1, s^2, \ldots$.
(2) variables for the terms of temporal coordinates $t, t^1, t^2, \ldots$.
(3) variables for the terms of conditional coordinates $u, u^1, u^2, \ldots$.
(4) variables for the terms of arbitrary coordinates and their combinations $v, v^1, v^2, \ldots$.

In addition to the definition of the propositional formula: if x is a propositional formula, and k is a variable for coordinates, then $[xk]$ is a propositional formula. In addition to the definition of the term form: if

a is a term form, and k is a variable for coordinates, then $[ak]$ is a term form.

Additional axiomatic schemes:

(I)

(1) $[[xt^1]t^2] \dashv\vdash [xt^1]$;
(2) $[[xs^1]s^2] \dashv\vdash [xs^1]$;
(3) $[[xs]t] \dashv\vdash [[xt]s]$;
(4) $[[xv^1]v^2] \dashv\vdash [xv^1v^2]$;

(II)

(1) $\vdash [[xt^1]t^2] \rightleftharpoons [xt^1]$;
(2) $\vdash [[xs^1]s^2] \rightleftharpoons [xs^1]$;
(3) $\vdash [[xs]t] \rightleftharpoons [[xt]s]$;

(III)

(1) $(\exists a)x \vdash (\exists v)[xv]$;
(2) $(\forall v)[xv] \vdash (\forall a)x$;
(3) $(\exists v)[xv] \vdash [\exists v]x$;
(4) $(\forall v)[xv] \vdash (\forall v)x$;

(IV)

(1) $[(x \wedge y)v] \dashv\vdash [xv] \wedge [yv]$;
(2) $[(x \vee y)v] \dashv\vdash [xv] \vee [yv]$;
(3) $[(x \to y)v] \dashv\vdash [xv] \to [yv]$;
(4) $[(\sim x)v] \dashv\vdash \sim [xv]$;
$[((\alpha \leftthreetimes a)x)v] \dashv\vdash (\alpha \leftthreetimes a)[xv]$;
$(\alpha \leftthreetimes a)[xv] \dashv\vdash (\alpha \leftthreetimes [av]x$;
$[\alpha \mathrm{P}(b)v] \dashv\vdash \alpha[\mathrm{P}v](b)$;
$[\alpha \mathrm{P}(b)v] \dashv\vdash \alpha \mathrm{P}([bv])$,

where α denotes the presence or the absence of $\urcorner$ or ?, $\leftthreetimes$ is either $\forall$ or $\exists$.

(V)

(1) $(u \to x) \vdash [xu]$;
(2) $(\exists \downarrow u)x \dashv\vdash (\exists s)(\exists t)x$;
(3) $(\forall \downarrow u)x \dashv\vdash (\forall s)(\forall t)x$.

Additional rules of inference:

(1) If $x \vdash y$, then $[xv] \vdash [yv]$;

(2) If $x \vdash y$, then $x \vdash y$ for all v;
(3) If $\vdash x$, then $\vdash x$ for all v.

Assertions

$$[\mathrm{P}v^1](a) \wedge \sim [\mathrm{P}v^2](a), [xv^1] \wedge \sim [xv^2],$$
$$(\forall[av^1])x \wedge \sim(\forall[av^2]x), \text{etc.}$$

are not contradictions, while the assertions

$$[\mathrm{P}v^1](a) \vee \sim [\mathrm{P}v^2](a), [xv^1] \vee \sim [xv^2],$$
$$(\forall[av^1])x \vee \sim(\forall[av^2])x), \text{etc.}$$

are not tautologies.

Let us adopt a definition: a statement x will be called local if and only if

$$(\exists v)x \wedge (\exists v) \sim x,$$

and universal, if and only if

$$(\forall v)x \vee (\forall v) \sim x.$$

If a statement is not local, this does not mean that it is universal, since

$$\sim((\exists v)x \wedge (\exists v) \sim x) \dashv\vdash ((\forall v)x \vee (\forall v) \sim x)$$
$$\vee ((?\forall v)x \vee (?\forall v) \sim x).$$

Similarly, the negation of the universality of a statement does not mean its locality, since

$$\sim((\forall v)x \vee (\forall v) \sim x) \dashv\vdash ((\exists v)x \wedge (\exists v) \sim x) \vee ((?\exists v)x$$
$$\wedge (?\exists v) \sim x).$$

Thus, there are instances when a statement can be classified neither as local nor as universal. Namely, it occurs when

$$((?\forall v)x \vee (?\forall v) \sim x) \wedge ((?\exists v)x \wedge (?\exists v) \sim x).$$

Thus, in this case we have uncertainty for the empirical sciences, which is excluded for the deductive sciences.

For conditions, there are rules, similar to those for spatial and temporal coordinates. The following rule

$$[xu^1u^2] \dashv\vdash [xu^1],$$

appears not to hold in this case. But it is valid without exception. This rule appears to be wrong because of its being confused with another statement

$$[x(u^1 \wedge u^2)] \dashv\vdash [xu^1],$$

which is indeed not valid in the general case (if u^2 is u^1 or if u^2 follows from u^1, this statement is correct).

It is appropriate to differentiate between conditions and coordinates from the point of view of the possibility of formulating new rules, not coinciding with (1)–(8). For instance, the following rule is possible:

$$\vdash (\exists a)x \rightarrow (\exists \downarrow u)(\forall a)\mathrm{M}(\downarrow [xu]),$$

where a is a variable for empirical individuals that occurs as free in x, and M is a predicate 'possible'. This rule may be used to obtain an inference. 'Under certain conditions any person may become the Emperor of Russia'. This conclusion seems to be a paradox. But if one tries to formulate the conditions under which certain people have become the emperors of Russia, the apparent paradoxicality disappears. It is even more so since we mean a possibility that is far from being always realized.

29. Consequences of Definitions

Consider a definition 'A diamond is an equilateral quadrangle'. The word 'is' plays here the role of symbol = Df., the determining part is the literature version of the term 'a quadrangle having all its sides equal'. It appears obvious that the assertions 'A diamond has all its sides equal' and 'A diamond is a rectangle' (where the word 'is' already plays a different role, namely, that of the predicate $\rightharpoondown$) are valid. These assertions represent the consequences of definitions.

However, the rules for obtaining the consequences of definitions have not been formulated explicitly, and it is impossible to explain why the above assertions are accepted without explicit formulation of these rules (incidentally, this is a good example of what has been said above concerning the role of logic in language processing).

The point is that as soon as the operation of the introduction of a new term b is completed (recorded, stated, read, heard, etc.), b becomes a term in its own right, and everything which has been said with respect to it, acquires an independent status. This fact may be recorded through the following rules:

(1) If one accepts $a = \mathrm{Df.}\,b$, then $a \rightleftharpoons b$ is regarded as true.

(2) If one accepts $x \equiv \mathrm{Df.}\,y$, then $x \equiv y$ is regarded as true.

Consider, for instance, a definition:

$$a = \text{Df.}\, b{\downarrow}\, \text{P}.$$

If it is accepted, then, by virtue of rule (1), the assertion

(1) $a \rightleftharpoons b{\downarrow}\, \text{P}$

is accepted as true. There is a logical rule, according to which

(2) $\text{P}(b{\downarrow}\, \text{P})$

is true, and another logical rule, according to which

(3) $\text{P}(b{\downarrow}\, \text{P}) \wedge (a \rightleftharpoons b{\downarrow}\, \text{P}) \rightarrow \text{P}(a)$.

As follows from (1) and (2),

(4) $\text{P}(b{\downarrow}\, \text{P}) \wedge (a \rightleftharpoons b{\downarrow}\, \text{P})$,

while from (3) and (4) we obtain

$$\text{P}(a).$$

30. Implicit Definitions

Definitions expressing a decision to use certain expression as a term or as a statement are called explicit. Such definitions deserve some merit, since their logical structure is obvious, and no additional requirements arise with respect to the terms introduced by them. But they are frequently quite cumbersome, and even unrealizable in practice. It is also somewhat difficult to obtain their necessary consequences. These shortcomings of explicit definitions are partially overcome by implicit definitions.

The implicit definitions are characterized by the following properties: (1) the newly introduced (defined) terms are accepted as fundamentally clear; (2) a certain set of statements is formulated as axioms; these statements contain the terms described in (1). The selection of the set of axioms is based on the following argument (although it is usually overlooked by the very authors of definitions): if the defined terms had been introduced through explicit definitions, these axioms would represent their consequences. The obvious convenience of this kind of definition is associated with deduction. Implicit definitions are especially convenient for introduction of the logical operators, and also in those cases when several terms and operators (or combinations of both) are introduced. Sometimes implicit definitions are used alongside with explicit.

For example, the term 'prime' may be implicitly defined by three axioms: (1) 'A prime is a positive integer'; (2) 'A prime can be divided only by one and itself'; (3) 'If a number is a positive integer and can be divided without a remainder only by one and itself, it represents a prime'.

Sometimes, in simple cases, implicit definitions easily turn into explicit, and vice versa. For instance, operator $\supset$ may be explicitly defined as

$$(x \supset y) \equiv \text{Df.} (\sim x \vee y),$$

or, in a different form, as '$x \supset y$ is an abbreviation for $\sim x \vee y$'. In its implicit form this operator is included in the alphabet of the system, and the set of axioms of this system is supplemented by those for $\supset$, in particular, by the following one:

$$(x \supset y) \dashv\vdash (\sim x \vee y).$$

But such transformations are not always practically possible and appropriate.

Implicit definitions have their own shortcomings. They do not express as in the explicit form that they are exactly definitions, introducing new terms or operators. This is why the authors begin to look for a meaning, additional with respect to what is expressed in the axioms containing them, and confuse the defining statements with those which also contain the defined terms but are accepted on different grounds.

Although in the case of implicit definitions, the latter take the form of axioms, i.e., assertions, this does not mean that the difference between the definitions and the axiom-statements disappears. Statements are constructed of terms the meaning of which is known before their construction and independently of them. Definitions, even in the case when they are given the form of axioms, contain terms the meaning of which is not known in the absence of these axioms.

31. Incomplete Definitions

There are implicit definitions which may be called incomplete, or mixed. They are constructed in the following manner. Situations are established in some way in which it is appropriate to use statements with a term *a*. There are no strictly defined criteria for these situations (since it is not necessary, practically impossible, nobody tried to do this, etc.). But such assertions as $x \vdash y$ or $\vdash (x \rightarrow y)$ are formulated (accepted), in which the defined term *a* occurs in *x* and does not occur in *y*. Since there are no reverse statements $y \vdash x$ and $\vdash (y \rightarrow x)$, it is impossible to eliminate statements with the term *a* from the language.

Let us give an example. It is known from linguistic practice what are the appropriate situations for using such expressions as '*a* generates *b*'. One may indicate certain criteria for such situations, in particular, the spatial proximity of *a* and *b*. But sometimes this ordinary usage and the following additional points of definition: (1) from '*a* generates *b*' follows '*a* appears before *b*', and (2) from '*a* generates *b*' follows 'if *a* does not appear, *b* does not appear', turn out to be sufficient. And only those points can later occur in the argument that claim logical rigour.

32. Pseudodefinitions

There are cases of introduction of linguistic expressions, which we call pseudodefinitions. They are constructed in the following manner. Consider a set of statements *x* introducing an expression *a*. Actually, *a* is generated as an abbreviation for *x* (this is usually overlooked), but later, after separating from the base that has produced it, *a* starts its life as an expression with independent meaning, regardless of the meaning of *x*. The peculiar feature of this process is that the expression *a* is given the form of expression P(*b*). But the purpose here is not to introduce the predicate P or the subject *b*. Instead, we are dealing with the linguistic practice of constructing statements which we divide into subjects and predicates, as well as with the totally unlimited possibility of constructing *a* from available linguistic elements. In such situations, the concrete form of these elements depends rather on the interests of observing the rules of logic in the introduction of terminology (we are repeating that these are usually completely overlooked), on most diversified factors – in the linguistic tradition, with a tendency towards literary expressiveness or sensation, etc. And if we add to the above that the expression *a* is introduced according to the scheme of implicit definition ('*a*, if and only if *x*'), then it appears that P(*b*) is perceived not as a result of a certain kind of linguistic creativity, but as an inference from a situation, described by a set of statements *x*. To put it briefly, on the surface it appears as if the statement P(*b*) follows from *x*.

To make the matters clear, let us first consider a common sense example. Assume that a person N, while building his career, made serious errors in the organization of the operation of his (her) department and in the relations with his (her) boss and colleagues. Assume that he (she) was laid off because of this, without a perspective to rise again. As is well known, in such situations the expressions 'N broke his (her) neck', 'N got broken', etc. are frequently used. These expressions represent abbreviated notations for a situation present for N, with notations understood and used under certain conditions and among a certain circle of persons. And nobody thinks of regarding the introduction of these expressions as definitions of the predicates 'broke his (her) neck', 'got broke', etc., or of interpreting them as statements in which these predicates are used literally.

One can find such examples in science also, the only difference being that here everything is understood literally and is interpreted as the result of the highest development of science, instead of the result of linguistic mystifications. Thus, the set of statements x concerning the behavior of a light beam in the vicinity of a body A becomes implicitly abbreviated by an expression 'The space in the vicinity of A is curved'. The latter is interpreted as a statement with a predicate 'curved', the meaning of which becomes separated from x and from the expression in which it appears in this case (everybody knows what a curved object is!). Similarly, a set of statements x concerning certain particles A under certain conditions B is implicitly abbreviated by an expression 'The time for A under conditions B is accelerated (slowed down)', and the words 'accelerated' and 'slowed down' are interpreted as independent predicates with the commonly known meaning, but in this case applicable to time.

Here we come across a special kind of linguistic mimicry, which may be noticed in nature and in many cases of linguistic practice of ordinary communication, but which is cultivated on a mass scale in a number of scientific fields as an index of progress of scientific thinking.

33. Operational Definitions

A special case of implicit definitions is represented by the definitions containing such expressions as $x \to (y \to z)$ and $(y \to z) \to x$, in which x is a statement with the defined term, y is a description of certain actions, z is a description of the result of these actions. For instance, the expression 'A weighs B kg' may be regarded as a replacement of the expression 'If A is put on a scale C, then the reading of the needle of C will be B'.

For a single defined statement x, different defining parts $y \to z$ are possible. Situations are also possible wherein the number of different $y \to z$ is not given, not limited, and in general not established through a common agreement. Thus, various versions are possible here, in particular the following ones:

(1) $x \leftrightarrow (y \rightarrow z)$;
(2) $x \leftrightarrow ((y^1 \rightarrow z^1) \vee \ldots \vee (y^n \rightarrow z^n))$;
(3) $((y^i \rightarrow z^i) \rightarrow x) \wedge (x \rightarrow (a \rightarrow b))$,

where $\downarrow y^i$ and $\downarrow z^i$ are arbitrary elements of certain specified classes, and a and b are the variables for the elements of these classes.

34. Intuitively Obvious Statements

There are instances when certain statements are accepted as intuitively clear or obvious. For example, 'An object, having a certain property, has this property', 'A straight line is the line of the shortest distance between two points'. Such statements represent either implicit definitions of the expressions occurring in them (operators, terms, parts of terms) or the consequences of such definitions. Thus, the first of the above statements is a part of the implicit definition of the limitation operator ('which') while the second is that of the implicit definition of the term 'straight line'. These assertions appear obvious not by virtue of proofs, observations, etc., but only because we ourselves accept them implicitly as definitions of the properties of the corresponding linguistic expressions. When the second of the above statements is rejected, the term 'straight line' is actually used with a different meaning, and this is all (by no means can we say that some new experimental facts are discovered, which contradict the earlier assumptions).

In this case intuition represents just the usage of linguistic expressions at the level of the elemental customs of their introduction. Since these customs are amorphous and their logical nature remains unknown, the statements under consideration become mystified. The problem of intuition arises, which appears subtle and complex, but which, under the conditions of a sufficiently thorough analysis of the logical rules for introduction of terminology, turns out to be fictitious. The identification (explication) of intuition is nothing but establishing the meaning of some linguistic expressions, occurring in the above statements, by the professionally developed methods of logic.

35. Variables

In order that the statements of logic acquire the desired generality, clear boundaries of their applicability, and a compact form, a special kind of symbols, called variables, is used. The variables are also employed in other sciences. In logic they are introduced for terms, statements, and logical operators.

The variables represent special objects that are similar to terms as far as the method of their formation and the ways of operating them are concerned. Ordinarily, these are letters or letter-like constructions. If they are regarded as empty spaces, the latter are differentiated and numbered in one or another way. Thus, the role of variables in such cases actually belongs to the symbols, differentiating and numbering these empty spaces.

The objects of the appropriate physical (perceivable) form are introduced as variables in two ways: (1) explicitly, or unconditionally; (2) implicitly, or conditionally. In the first case one simply enumerates the objects that will become variables of some kind. For instance, a decision is made to consider the letter a with indices $1, 2, 3, \ldots$ as variables for statements. In the second case the variables are not enumerated. Their role is taken by arbitrary objects of appropriate form, with respect to which it appears clear from the context what kind of variables they represent. For instance, in the expression 'If a and b are statements, then $a \wedge b$ is a statement' the letters a and b play the role of variables for statements; this can be seen from the expression itself. In science (and not just in logic and mathematics) not infrequently variables are introduced implicitly. The convenience of the first method is that the variables are clearly fixed. The second way is convenient inasmuch as the same objects may serve as variables of different kinds (sometimes it is not so simple to invent variables).

In place of variables one can substitute certain objects that form the set of values for variables. An important factor here is that these objects themselves represent the elements of language – terms, statements, parts of terms, operators, etc. These are the objects that can indeed be substituted in place of variables. When the set of values for variables is specified as that of certain objects having nothing to do with the language, this is due either to custom (forget the logical essence of the variables!) or to conveniences of the language (we imply, for instance, that in place of variables for contacts one substitutes not the electrical circuit contacts themselves, but their notations) or to the existence of certain conceptions, etc.

The objects, substituted in place of variables, may be divided into two groups. This division does not mean that in all cases of variable use it is preserved for the same variables. The first group includes those objects that occur in the language of a given construction (theory, calculus) or may be formed of the linguistic elements of this construction. For instance, such is the substitution of propositional formulae (including propositional variables) in place of propositional variables. This kind of substitution represents the technical means for obtaining theorems within the framework of given constructions. We shall call them formal, or syntactical. In such cases we shall speak of the syntactical set of values for variables.

The second group includes those objects that do not appear in the language of construction (theory, calculus, or, in general, a certain fragment of science) with given variables and cannot be constructed in it. These are, for instance, the substitutions of sentences of concrete real languages in place of propositional variables. This kind of substitution we shall call the interpretational, or semantical. We shall speak in such cases of the semantic set of values for variables (or of the domain of their interpretation, or application). It is only through interpretational substitutions that one obtains statements from expressions with variables.

The relation between variables and objects from the semantic set of their values is similar to that between terms and the objects denoted by them. We must recall that in this case the object set is formed by terms, statements, their parts, operators, etc., or, to put it more briefly, the elements of the language. Thus, the relation between the propositional variable *a* and the statement *b* is the relation between *a* as a term and *b* as an object denoted by this term.

The substitution of objects from the set of values of variables in place of variables is governed by the following rules:

(V^1) In place of any variables of a given kind, considered separately, any object from the set of its values may be substituted.

(V^2) In place of different occurrences of the same variables in a given expression the identical objects (but not the same object) from the set of values of this kind of variables may be substituted.

It is important to point out here that if a term *b* is substituted in place of two different occurrences of a variable *a* in a given expression, the different occurrences of *b* in the resulting expression do not always denote the same object. They may denote different individuals of the same kind.

(V^3) In place of different variables in a given expression different objects from the set of their values may be substituted.

(V^4) In the case of semantic substitution the substitution of an object *b* in place of a variable *a* is performed in all the places where *a* occurs in a given expression.

Thus, one may regard the variables as terms, but as a special kind of term. The peculiar feature of the variables is that two or more objects must be fixed, and it must be taken into account that the objects of a given kind may be arbitrary, that they may be differentiated in some way (but it does not make a difference in what way exactly), etc. For example, the variables for terms denoting tables (it is more convenient to say 'the variables for tables') may be distinguished as expressions: 'one table', 'another table', 'the third table' (the 'third' meaning not the order, but the difference from the first two), etc. Each of the above expressions denotes an arbitrary table. But if they appear together, the differences among them mean that if one of these expressions denotes a certain table, then the other ones may denote different tables. In other words, the variables are but technical means simplifying the usage of terms in situations with sets of objects.

On the other hand, not every general term may be regarded, in its turn, as a variable. Consider, for example, the term 'number' and the species term 'odd number'. The first one will be considered as a variable, the second as an object from the set of its values. The statement 'Certain numbers may be divided by two without a remainder' is true, while the statement 'Certain odd numbers may be divided by two without a remainder', obtained from the first one by the rule (V^4), is false.

Let us adopt the following rules for variables representing the variables for terms:

(V^5) From the statement 'If $a^1, \ldots, a^n$ ($n \geqslant 1$) are b, then A' follows the statement 'If $a^1, \ldots, a^{n-1}$ are b, then $(\exists q)$B', where B is formed from A through substitution of q in place of one or more free occurrences of a^n in A, and q is a variable for b that does not occur in A.

If $n = 1$, then (V^5) takes the form: from 'If a is b, then A' follows $(\exists q)$B. For example, from the statement 'If a and $a + 1$ are positive integers, then $a + 1 > a$' follows 'If a is an integer (positive), then $(\exists q)(q > a)$', where q is a variable for positive integers.

(V^6) From the statement 'If $a^1, \ldots, a^n$ are b, then A' follows 'If $a^1, \ldots, a^{n-1}$ are b, then $(\forall q)$B', where B is formed of A through substitution of q in place of all free occurrences of a^n in A, and q is a variable for b that does not occur in A.

For $n = 1$ (V^6) takes the form: from 'If a is b, then A' follows $(\forall q)$B. An example for (V^6): from 'If a is a positive integer, then $(\exists q)(q > a)$' follows $(\forall r)(\exists q)(q > r)$, where r is a variable for positive integers.

36. Definitions with Variables

A special case of definitions is represented by definitions with variables. They are constructed according to the following logical scheme: if $a^1, \ldots, a^n$ ($n \geqslant 1$) are terms such that the statements x with these terms are true, then $b^1, \ldots, b^m$ ($m \geqslant 1$) will be terms such that statements y with these terms are true. Here $a^1, \ldots, a^n$ play the role of variables for terms.

The rule for definitions with variables: in the definition itself as well as in its consequences one should not substitute $b^1, \ldots, b^m$ and all those terms that either contain $b^1, \ldots, b^m$ or are defined through them in place of $a^1, \ldots, a^n$. This rule is nothing but the identification of the condition, present in the definition itself, according to which $a^1, \ldots, a^n$ should represent terms regardless of the definition of $b^1, \ldots, b^m$.

Violation of the above rule may lead to paradoxical consequences. It is exactly in this way that the well-known paradox of the set of normal sets is produced.

For the sake of greater visual clarity, let us adopt the abbreviations: n – a normal set, Mn – a set of normal sets, $\in$ – the symbol of inclusion of an element in a set ($a \in b$ reads as 'a is an element of b', 'a is included in b', 'a is contained in b', etc.). The definition of a normal set may be given as follows: n is a term such that if a is a term of a set, then

$$\sim (a \in a) \rightarrow (a \in \mathrm{M}n),$$
$$(a \in a) \rightarrow \sim (a \in \mathrm{M}n),$$

(in other words, if a is not included in itself, then a is included in the set of normal sets).

The paradox results if one substitutes in place of the variable a the term Mn:

$$\sim (\mathrm{M}n \in \mathrm{M}n) \rightarrow (\mathrm{M}n \in \mathrm{M}n),$$
$$(\mathrm{M}n \in \mathrm{M}n) \rightarrow \sim (\mathrm{M}n \in \mathrm{M}n),$$

(in other words, if Mn is not included in Mn, then Mn is included in Mn, and if Mn is included in Mn, then Mn is not included in Mn). This substitution appears legitimate, since Mn is a term of a set. But it is wrong, since, according to the rules for definition, the expression Mn as well as the expression n containing it do not represent terms at all before the construction of the definition, and according to the very formulation of the definition, a is a variable for those objects that are terms before the construction of the definition and independently of it (and only for those objects that represent terms denoting the sets). Thus, if one strictly observes the rules for definition, the paradox does not result.

Using variables, one may construct the following kinds of implicit definitions for one term:

(1) $(a \rightharpoonup b) \rightarrow (x \dashv\vdash y)$;
(2) $(a \rightharpoonup b) \rightarrow (x \leftrightarrow y)$;
(3) $(a \rightharpoonup b) \rightarrow (\vdash x)$;
(4) $(a \rightharpoonup b) \rightarrow (x \vdash y)$;
(5) $(a \rightharpoonup b) \rightarrow (x \rightarrow y)$,

and also the Definitions (6)–(10), in which the symbol $\leftrightarrow$ appears instead of $\rightarrow$ (i.e., the reverse statement is valid). In (1)–(10) the letter a is a variable, and b is the newly introduced term. Definitions (4), (5), (9), (10) are incomplete (or partial) definitions, since the question of acceptance of $y \vdash x$ or $y \rightarrow x$ remains open or it is known that these assertions are not true.

37. The Multiple Meanings of Linguistic Expressions

The negative role of the multiple meanings of linguistic expressions in science and, in general, in the linguistic practice of humans is well known. And fighting it is all the more complex because of it being not always noticeable; sometimes its detection requires a thorough analysis. First, the sets of objects, denoted by the given terms, may nearly coincide, so that an insignificant difference between them is completely neglected (although it is exactly this difference that can play a fateful role). Second, it is possible that one may imply the same set of objects, but the terms denoting them are introduced by different methods, so that different consequences naturally arise. The above is also true for logical operators.

Consider, for instance, the term 'space'. From the common usage of words it appears clear what is meant when the word 'space' is mentioned. But in some cases one has in mind spatial volumes, regardless of the objects that are situated in or that may occupy them, while in other cases it is volumes with all the objects contained in them or with a special substance filling these volumes. Although here we come across simply a different word usage, one group of persons accuses the other of making errors. The modern 'theoretical' spheres of science are truly abundant with such situations.

38. Explication

There are terms and statements the meanings of which are considered clear from the point of view of certain customary word usage, and more or less unambiguous. In this case the terms are not required to be strictly defined, and the statements are not required explicitly to express their logical structure. Moreover, sometimes the rigorous definition of terms as well as the construction of sentences according to the standards of logic is in general difficult or practically impossible. And if there arises a need to employ such terms and statements in a certain science, their ordinary usage may prove unacceptable: it is characterized by multiple meanings, making it difficult or completely impossible to obtain necessary consequences, does not guarantee the correctness of the argument, etc. Therefore, for certain purposes of science, instead of these terms and statements, a kind of substitute is used, which is strictly defined, and the logical structure of which is expressed in a sufficiently explicit form. This operation of introducing terms and statements, replacing certain given terms and statements, is called the explication of the latter.

The task facing the explication of linguistic expressions consists in defining or giving a clear formulation of some expressions that permit the application of logical rules and are convenient from the point of view of deducing the necessary conclusions guaranteeing the unambiguous nature of the verification results, etc.; these expressions at the same time coincide in a sense with the given linguistic expressions. One should not consider the enumeration of the various meanings of linguistic expressions or the selection of one case of their usage as the task of explication. If in the process of explication the same words and formulations are preserved, still 'duplication' of terminology takes place.

The form of explication of linguistic expressions, based on the usage of just the means of logic and certain common habits of linguistic practice, will be called 'logical explication'. The above habits represent the ordinary cultural element and do not require the knowledge of any special science for their understanding. In the above discussion we have made use of logical explication in a systematic manner; in particular, we have introduced the operators $\sim$, $\vee$, $\wedge$, etc. as expressions explicating the well-known linguistic expressions 'not', 'or', 'and', and so on. In what follows explication will be also the most important means of our analysis.

Explications may be compared. Let $a^1 = \text{Df.}\, b$ and $a^2 = \text{Df.}\, c$ be the explications of the same expression a. In a similar way, assume that $a^1 \equiv$ Df. b and $a^2 \equiv$ Df. b. The following situations are possible: (1) $b \rightharpoonup c$ (in the case of definitions with $\equiv$ Df.); (2) $(c \rightharpoonup b) \wedge \sim(b \rightharpoonup c)$ or, respectively, $(c \rightarrow b) \wedge \sim(b \rightarrow c)$; (3) $b \rightleftharpoons c$ or, respectively, $b \leftrightarrow c$; (4) $\sim(b \rightharpoonup c) \wedge \wedge (c \rightharpoonup b)$ and, respectively, $\sim(b \rightarrow c) \wedge \sim(c \rightarrow b)$. Thus, there may be situations when people speak of the same objects but cannot understand each other because of different ways of definition of the same expressions.

39. Consistency of Terms

Terms may be constructed or defined in such a way that from statements containing them contradictory expressions may be deduced. For example, from the statement $(\wedge\, P\tilde{P})(a)$ one may deduce $P(a) \wedge \sim P(a)$, and the cause of this lies in the structure of the predicate; if the definition $a =$ Df. $b{\downarrow}(\wedge\, P\tilde{P})$ is accepted, then $P(a) \wedge \sim P(a)$ will be provable, and the cause of this is the above definition.

We shall consider a subject a to be contradictory if and only if there is such a simple and undefinable predicate P, that $P(a) \wedge \sim P(a)$ is provable. The predicate P is considered to be contradictory if and only if there is such a simple and undefinable subject a that $P(a) \rightarrow Q(a) \wedge \sim Q(b)$ is provable. Terms are consistent if they are not contradictory.

CHAPTER TWO

THE SPECIAL THEORY OF INFERENCE AND TERMS

1. The Logical Explication of Terms

The task of logic is not just to study the general properties of terms and expressions, associated exclusively with the fact that they represent terms and expressions, and with the logical operators and the combinations of logical operators, terms and statements contained in them. It also studies various terms and expressions through establishing their meaning, or by realizing their explication by the methods of the general theory of inference and terms. Examples of such expressions are 'individual', 'class', 'cluster', 'exists', 'possible', etc. From this point of view, logic represents the general semantics of science.

2. Empirical and Abstract Objects

It is only in the final chapter that the distinction between empirical and abstract objects will be of significant importance. But since many of the definitions accepted in this chapter do not agree with the usual definitions of the corresponding expressions that have been formed precisely in connection with abstract objects and that are not applicable to empirical objects, we shall, in order to give an explanation, have to take into account the difference between abstract and empirical objects. Therefore, in any case, we shall offer a few remarks here concerning this issue.

The objects that are perceived, sensed, observed, etc. by the investigator, in other words, the objects that act on the scientist's sensory apparatus, are empirical objects. They exist at a certain time and in certain areas of space; they appear, vary and become destroyed, influence each other, etc. To put it briefly, they include everything that may be said in general concerning the research activities of people

The investigator may make a decision not to take into account certain characteristics of the objects under study, or to take into consideration only some of their characteristics. This decision may be carried out by selection of the appropriate research situation or through an artificial creation of such situation. In this case, however, the objects under study remain empirical objects, but considered under special conditions.

The situation is quite different if one adopts a decision to neglect those characteristics of object without which the empirical objects cannot exist. For instance, the investigator may decide to ignore the dimensions and shape of physical bodies when considering their motion, under the assumption that these bodies have no spatial dimensions (they represent 'material points'). Such a decision may be realized by the assumption of the existence of special objects that are called abstract, or ideal. This kind of object does not exist empirically, by the very nature of the underlying assumption. And study of such objects will not represent a process of observation. Instead it will be a process of determination of the appropriate definitions and axioms (postulates) and of deducing necessary conclusions from them.

The terms for abstract objects are introduced, first of all, through definitions, which may be presented explicitly and schematically as follows: (1) we shall call by a term *a* those objects that have the attributes $P^1, \ldots, P^n$ $(n \geqslant 1)$ but do not have the attributes $Q^1, \ldots, Q^m$ $(m \geqslant 1)$; the attributes $Q^1, \ldots, Q^m$ are such that if the empirical objects have the attributes $P^1, \ldots, P^n$, they also have the attributes $Q^1, \ldots, Q^m$; (2) we shall call by a term *a* those objects that have the attributes $P^1, \ldots, P^n$, and if from the fact that they have these attributes, and from other assertions accepted in a given science, it does not follow logically that $Q(a)$, then we have $\neg Q(a)$; the attribute Q may be such that if an empirical object has attributes $P^1, \ldots, P^n$, it also has Q. Then the terms for abstract objects are introduced according to the general rules for construction of terminology with the terms, introduced by Definitions (1) and (2), used as the initial base.

Examples of terms for abstract objects: the terms 'material point', 'ideal gas', 'incompressible fluid', etc. In the case of a material point it is assumed that a physical body has mass but no spatial dimensions.

The objects denoted by terms for abstract objects do not exist empirically according to the very conditions for construction of these terms. It does not follow from this that these terms do not denote anything. Thus, the term 'material point' denotes physical bodies possessing mass but not possessing spatial dimensions. As far as the relation of these terms to empirical objects is concerned, it is governed by the following rule: a decision is taken to consider certain empirical objects *a* such that $a \rightharpoonup b$, where *b* is a term for abstract object. For instance, a decision is taken to consider the planets of the Solar system and the Sun as material points.

It is an extralogical question, how the objects a are selected. In this case it is important that the relation $a \rightharpoonup b$ is accepted according to the investigator's choice. Then from $a \rightharpoonup b$ and from certain statements containing b, one can obtain, as logical conclusions, the statements containing a. The objects a are selected in such a way that these latter statements could be accepted as true. Deviations from the desirable result presented here do not change the essence of the operation.

In practice the situation is as follows: there are known empirical objects a, having the attributes $P^1, \ldots, P^n$; it is known that their attributes Q^i may be neglected; and the term for the abstract object b is introduced in such a way that a decision is taken beforehand to accept $a \rightharpoonup b$; but the set of objects that may be regarded as objects b is not limited to the objects a; in principle it is not bounded, and any objects, similar to objects a (from the point of view of the above attributes), may be regarded as objects b.

The selection of the empirical object a such that $a \rightharpoonup b$ represents the empirical interpretation of the abstract object. If the terms for abstract objects are complex, and there are derived predicates introduced for them, then the empirical interpretation of abstract objects proves to be a rather subtle and not always possible process. And a whole number of methodological difficulties results from violation of the rules for such interpretation.

If b is a term for abstract object, and P is a predicate, introduced for the terms of abstract objects and appropriate with respect to b, then in the empirical interpretation of b the problem of the possibility of expanding the sphere of application of the predicate P over a should be solved in a specific way in each case. It is not excluded that this interpretation of b will represent the interpretation of P(b), for which one selects for the object a a certain predicate Q such that $\downarrow Q(a) \rightharpoonup \downarrow P(b)$.

3. Individuals

The subject-term that cannot be generic with respect to an arbitrary term will be called individual, while the objects denoted by them, the individuals. In other words, a subject-term a is individual if and only if for any term b the following is valid: if $b \rightharpoonup a$, then $a \rightharpoonup b$. Examples of individual terms: 'the first astronaut, having performed an orbital flight around the Earth', 'a Russian poet M. Yu. Lermontov, killed in a duel in 1841', 'a planet of the Solar system, Earth', etc.

Let a and b be the variables for subject-terms. The above definition may be given the following form: b represents an individual term if and only if

$$(\forall a)((a \rightharpoonup b) \rightarrow (b \rightharpoonup a)).$$

It follows from the definition: if a is an individual term, and $b \rightharpoonup a$, then $a \rightharpoonup b$ and $a \rightleftharpoons b$.

The number of objects denoted by the individual term equals one; therefore these terms are also called unitary. But this characteristic is extra-logical, not promising from the point of view of obtaining logical conclusions.

We shall call the variables for individual terms, individual variables.

We shall use such expressions as 'An individual *a*' and '*a* is an individual'. The first of them is but a linguistic transformation of the term 'An object, denoted by the individual term *a*'. The second one is a linguistic transformation of the statement 'A term *a* is individual'. On the other hand, the expression 'individual' itself (as a subject-term) coincides with the term 'object'. When it is used with a specific meaning, it is done with the intention to stress the irreproducible nature of the object. But regardless of our explanation, logical rigor here depends only on reference to the individual quality of the term. Whether we say 'An individual *a*', '*a* is an individual', 'Consider an individual *a*', 'Let *a* be an individual', etc., the exact meaning of all these statements will be the same: *a* is an individual term, in other words, for any term *b*, if $b \rightharpoonup a$, then $a \rightharpoonup b$.

In addition to the rule provided by the definition, for individual terms the following rules are valid: if *a* is an individual term, then

(1) $x \vdash (\forall a)x$;
(2) $(\exists a)x \vdash x$.

Thus, for individual terms quantifiers are superfluous.

4. Classes (Sets)

We use the expressions 'class' and 'set' as synonyms. Further, we distinguish between the logical and mathematical theories of classes (sets). The task of the logical theory of classes is to establish such rules for operating with the terms of classes, and statements with these terms, that they would not depend on the concrete properties of classes.

We draw a difference between the term 'class' and a special term-forming operator which also reads as the word 'class' but does not represent a term. As such an operator, the symbol K will be used.

By means of operator K the terms of primary classes are formed according to the following rule: if a is a subject-term, then Ka is a subject-term and at the same time Ka is an individual term.

To form a primary class means to construct a term Ka, in other words, to say literally 'a class of objects a'. For instance, to form a class of gods means to form a term 'class of gods', where the word 'class' is the operator K; to form a class of microparticles means to construct a term 'a class of microparticles', etc.

The terms for primary classes form the basis for the whole terminology denoting classes. In particular, the term 'class' is introduced through generalization of the terms of classes. Namely, it may be done in the following way (by making use of the variable for the terms of classes): the word 'class' will be a term such that if a is a term of class, then $a \rightharpoonup$ 'class'. With the aid of the operation of limitation, one can introduce such terms as 'a class such that P' (for instance, 'an empty class') and 'a class such that x' (for example, 'a class such that all its elements have attributes $Q^1, \ldots, Q^n$').

Statements to the effect that the objects denoted by the term a, are included in a class denoted by the term B, will be symbolically written in the form

$$a \in \mathrm{B}.$$

The predicate of inclusion of individuals in a class (i.e., $\in$), appearing in the above expression, is defined simultaneously with the properties of class-forming operators, in particular, simultaneously with K. The individuals, denoted by the term a, represent the elements of class B. But the term a may be general.

The properties of terms with operator K and the predicate $\in$ are defined implicitly by a system of assertions that may include the following ones:

$$(\exists a)(a \in \mathrm{K}b) \vdash (\exists b)(b \in \mathrm{K}a);$$
$$(\forall a)(a \in \mathrm{K}b) \wedge (\forall b)(b \in \mathrm{K}c) \vdash (\forall a)(a \in \mathrm{K}c);$$
$$(\exists a)(a \in \mathrm{K}b) \wedge (\forall b)(b \in \mathrm{K}c) \vdash (\exists a)(a \in \mathrm{K}c);$$
$$(a \in \mathrm{K}b) \wedge (a \in \mathrm{K}c) \vdash (\exists b)(b \in \mathrm{K}c).$$

Here as well as in a number of cases below we do not list the complete system of axioms, since it will not be necessary. And in order to illustrate the essence of the matter, it would suffice to give individual examples.

By making use of the predicate $\in$ and operator K, one may also define the term 'class' (in abbreviated form, *kl*) in the following manner: if a is a class-term then $a \in Kkl$. Obviously, if b is a subject-term, then Kb is a class-term, and therefore, $Kb \in Kkl$.

A statement to the effect that the class A is included in the class B will be written in symbolic form as

$$A \subset B.$$

The predicate of inclusion of a class in a class (i.e., $\subset$), appearing in the above expression, is defined through the predicate $\in$ implicitly in the following manner:

$$(\forall a)((a \in A) \rightarrow (a \in B)) \dashv\vdash (A \subset B)$$
$$(\exists a)(((a \in A) \wedge \sim(a \in B)) \vee ((a \in B) \wedge$$
$$\wedge \sim(a \in A))) \dashv\vdash \sim(A \subset B)$$
$$(\forall a)(a \in A) \dashv\vdash (Ka \subset A),$$

where a is an individual variable.

One can form the same kind of terms of class-terms by using the operators of unification ($\cup$), intersection ($\cap$) and complementation ($-$) of classes. The rules for constructing such terms: (1) if a and b are class-terms, then $a \cup b$ and $a \cap b$ are class-terms; (2) if a is a class-term, then $\bar{a}$ is a class-term. The properties of these operators are specified, in particular, by these assertions:

$$(a \in A) \wedge (a \in B) \dashv\vdash (a \in A \cap B);$$
$$(a \in A) \vee (a \in B) \dashv\vdash (a \in A \vee B);$$
$$(a \in B) \dashv\vdash (a \in \bar{\bar{B}}).$$

For the existence of the primary class, it is sufficient that its term is constructed. Thus, having formed the expression ‘a class of gods’, we have formed a class of gods, and it has started its existence regardless of whether gods do exist or not. The problem of the existence of derived classes is solved, by depending on the logical rules observed during the construction of their names and on additional definitions. In particular, if from the definition of the class-term it follows that $(a \in A) \wedge \sim(a \in A)$, then A does not exist.

The existence of classes does not in general depend on the existence of individuals included in them. Thus, an integer which is equal to the square root of five does not exist, while the class of integers equal to the square root of five, does exist. In particular, it may be said to be empty (i.e., there is no such integer that may be included in it without making an error). And vice versa, the existence of individuals does not depend on whether they are included in certain classes or not, and on what classes they are included in. The class-term may be constructed in such a way that this class will clearly not exist. But this class may include existing individuals. For instance, let us form the following class-term: 'a class, which includes an object a and at the same time does not include it, and which also includes electrons'. While this class does not exist, since in its formation we have violated the rules of logic, the electrons, as is well known, do exist.

Thus, one can see that classes are objects such that their existence depends exclusively on the construction of their names. And when attempts are made at defining classes as something that exists independently of their terms, classes become confused with n-tuples and clusters of objects. This confusion is not always indifferent. For example, the triplet of integers such that the sum of the third power of two of them equals the third power of the third one does not exist; at the same time the class of such triplets does exist; in particular, one may say about it that it is empty. But it is impossible to visualize such a class as something existing independently of the class name.

The study of classes is the study of objects included in these classes. Therefore, all the predicates for classes are defined with reference to their elements. Let us give a few examples.

A class A is a subclass of a class B if and only if $A \subset B$. And since $A \subset B$ is but an abbreviation for the expression $(\forall a)((a \in A) \rightarrow (a \in B))$, the first expression in the above definition of the subclass can be replaced by the second.

Classes A and B do not intersect if $(\forall a) \sim ((a \in A) \wedge (a \in B))$, and do intersect if $(\exists a)((a \in A) \wedge (a \in B))$, where a is the individual variable.

A class A is empty if $(\forall a) \sim (a \in A)$, and non-empty if $(\exists a)(a \in A)$. A class A is universal if $(\forall a)(a \in A)$, and non-universal if $(\exists a) \sim (a \in A)$.

Classes A and B are equivalent if and only if $(A \subset B) \wedge (B \subset A)$ or, in a different form, if $(\forall a)((a \in A) \leftrightarrow (a \in B))$.

A class is regarded as finite if the number of its elements is finite, and infinite if the number of its elements is infinite. The expression 'cardinality' of a class' is just a substitute for the expression 'the number of elements in a class'. All predicates of order for classes are defined by reference to ordering of their elements.

5. Clusters

As in the case with classes, we shall distinguish between the operator 'cluster' and the term 'cluster'. The latter is defined (in analogy with the term 'class') in the following way: if *a* is a term, denoting a cluster of individuals, then $a \rightharpoonup$'cluster'.

The primary terms of clusters are constructed according to the rule: if *a* is a subject-term, then 'cluster *a*' is a subject-term. The other terms for clusters of objects are formed according to the rules for construction of terminology. Note that unlike the term 'class *a*', the term 'cluster *a*' is not necessarily individual.

An example of the difference between the operator and the term 'cluster': in the term 'a cluster, consisting of *a*, *b* and *c*' the word 'cluster' is a term, while in the term 'a cluster of stars in the region A' it is an operator. When such expressions as 'houses situated in the area A', 'molecules in a given volume of gas', 'stars, included in the Galaxy', etc. are used, it is frequently not the classes, but the clusters of the corresponding objects that are implied. This is an approach to objects which is different from that in the case of formation of classes. Thus, as far as class is concerned, it would be meaningless to speak of spatial dimensions, motion, etc., while such predicates are quite appropriate in the case of clusters. The existence of a class does not depend on the existence of the individuals included in it, whereas the existence of clusters does depend on it. Thus, a cluster of *a* and *b* exists if and only if each of *a* and *b* does exist, whereas a class, including *a* and *b* exists if the term 'class, including *a* and *b*' is formed.

As in the case with classes, we shall denote by such symbols as $a \in B$ the fact that an object *a* represents an element of (is included in) a cluster B. Symbols of the form $A \subset B$ will mean (as in the case of classes) that all objects, included in cluster A, are included in cluster B.

If the term of cluster A is constructed in such a way that with respect to any individual *a* it is known whether it is included in A or not, then A is an individual term, and the cluster denoted by it is an individual.

Such expressions for clusters as $B \cup C$, $B \cap C$, $\bar{B}$ will be regarded, respectively, as the join (union) of B and C, as the common part (intersection) of B and C, and as a cluster of individuals, not included in B.

In the case of clusters, a system of definitions may be constructed which is partially similar to definitions for classes and partially different from it. In particular, the following assertions are accepted for clusters, that do not hold for classes:

(1) $\vdash Et(A) \rightarrow ((b \in A) \rightarrow Et(b))$;

(2) $\vdash Es(A) \rightarrow ((b \in A) \rightarrow Es(b))$,

where A is the variable for clusters, *b* is the variable for empirical individuals, Et is the predicate 'Exists at a time t', Es is the predicate 'Exists in the spatial domain s'.

6. States, Events

Let x be a statement $\alpha P(a^1, \ldots, a^n)$, where $n \geqslant 1$, and α denotes the presence or the absence of quantifiers, negations and the uncertainty operator in various admissible combinations or a combination of such statements constructed by means of statement-forming operators. In this case $\downarrow x$ is a term of state. The objects, denoted by such terms, are states.

A term $a^i (i = 1, \ldots, n)$ will be called free in x if α does not contain the corresponding quantifier $(\forall a^i)$ and $(\exists a^i)$. The term $\downarrow x$ will be considered individual (and a state, denoted by it, the individual state) if and only if all free terms among $a^1, \ldots, a^n$ are individual terms.

Such expression as 'event' is also used. It is usually associated with those $\downarrow x$ in which the predicate of statement x contains an indication of the fact that x became true, while before $\sim x$ was true. For instance, the term 'the fact that a particle a entered a domain b' denotes an event. Every state corresponds to events.

7. Existence

The predicate 'exists' will be written in abbreviated form as a symbol E. The usage of such a predicate is an empirically established fact, since there are statements like 'The electron exists', 'Phlogiston does not exist', 'Christ as a historic personality did not exist', etc.

Objects either exist or do not exist in a certain place, in a certain time, in any place, at any time, etc. i.e. in a certain reference frame. Although coordinates will be implied in considering the predicate E, we shall ignore them for the sake of simplifying the discussion (they do not characterize E specifically). And, incidentally, the introduction of the symbol E in this situation carries out still another task: to abstract oneself from those associations that arise in connection with the grammar form of the word 'exists'.

There are cases when the meaning of the term E, instead of being defined, is just explained. Those are the cases of usage of E in statements E(a), $\neg$E(a) and ?E(a), where a is an individual term. There are two subcases here: (1) the problem of the existence of individuals which is solved by depending on the possibility of observation of them by means of the sensory organs or by means of measuring devices, directly or from their traces (from their action on other objects); trust in the evidence of those who observed one or another individual is included in this group; (2) the problem of the existence of individuals is solved by means of deriving proof from other data, by making an assumption for certain purposes, by making an inference from definitions, etc. The uncertainty in this situation means that it is impossible to establish either through observation or through argument whether the individual does or does not exist. For example, in the case of the process of creation or annihilation of an individual.

Assuming that the value of E is given for the above cases, one can construct an exact definition of this predicate for other cases.

Suppose that a in statements $E(a)$, $\urcorner E(a)$ and $?E(a)$ may be a generic term. For instance, in the expression 'The electron exists' the word 'electron' is a common term. Definitions of E for common subjects have the following form:

$$E(a) \equiv \text{Df. } (\exists b)((b \rightharpoonup a)\, E(b)),$$

where b is the individual variable, or, to put it differently,

$$E(a) \equiv \text{Df. } (\exists b)\, E(b \downarrow (b \rightharpoonup a)),$$

where reads as 'a exists if and only if there is at least one individual that is a';

$$\urcorner E(a) \equiv \text{Df. } (\exists b)((b \rightharpoonup a) \to \urcorner E(b)),$$

or, to put it differently,

$$\urcorner E(a) \equiv \text{Df. } (\forall b) \urcorner E(b \downarrow (b \rightharpoonup a)),$$

which reads as 'a does not exist if and only if all the individuals that are a, do not exist'.

$$?E(a) \equiv \text{Df. } \sim E(a) \wedge \sim \urcorner E(a).$$

In the case of n-tuples of subjects the predicate E is defined in the following manner:

$$E(a^1, \ldots, a^n) \equiv \text{Df. } E(a^1) \wedge \ldots \wedge E(a^n),$$

in other words, the n-tuple of objects does exist if and only if every object, included in it, does exist;

$$\urcorner E(a^1, \ldots, a^n) \equiv \text{Df. } \urcorner E(a^1) \vee \ldots \vee \urcorner E(a^n),$$

i.e., the n-tuple of objects does not exist if and only if at least one of the objects included in it does not exist;

$$?\mathrm{E}(a^1,\ldots,a^n) \equiv \mathrm{Df.} \sim \mathrm{E}(a^1,\ldots,a^n) \wedge \neg\mathrm{E}(a^1,\ldots,a^n).$$

In the implicit form the above definitions appear as follows:

(1) $\mathrm{E}(a) \dashv\vdash (\exists b)\mathrm{E}(b{\downarrow}(b \rightharpoonup a))$;
(2) $\neg\mathrm{E}(a) \dashv\vdash (\forall b)\neg\mathrm{E}(b{\downarrow}(b \rightharpoonup a))$;
(3) $?\mathrm{E}(a) \dashv\vdash \sim \mathrm{E}(a) \wedge \sim \neg E(a)$;
(4) $\mathrm{E}(a^1,\ldots,a^n) \dashv\vdash \mathrm{E}(a^1) \wedge \ldots \wedge \mathrm{E}(a^n)$;
(5) $\neg\mathrm{E}(a^1,\ldots,a^n) \dashv\vdash \neg\mathrm{E}(a^1) \vee \ldots \vee \neg\mathrm{E}(a^n)$;
(6) $?\mathrm{E}(a^1,\ldots,a^n) \dashv\vdash \sim \mathrm{E}(a^1,\ldots,a^n) \wedge \sim \neg\mathrm{E}(a^1,\ldots,a^n)$.

The existence of the individual states is determined by the axioms:

(7) $\mathrm{E}({\downarrow} x) \dashv\vdash x$,

i.e., ${\downarrow} x$ does exist if and only if x is true.

(8) $\neg\mathrm{E}({\downarrow} x) \dashv\vdash \sim x$,

i.e., ${\downarrow} x$ does not exist if and only if $\sim x$ is true. If ${\downarrow} x$ represents a generic term, then the predicate of existence is defined for ${\downarrow} x$ as in the general case (see above). Since either x is true or $\sim x$ is true, for the existence of states the negations $\sim$ and $\neg$ will coincide; this may be expressed by the axioms:

(9) $\sim \mathrm{E}({\downarrow} x) \dashv\vdash \neg\mathrm{E}({\downarrow} x)$.

From the above axioms one can obtain the conclusions, for instance, such as

$$\mathrm{E}({\downarrow}(x \wedge y)) \vdash \mathrm{E}({\downarrow} x) \wedge \mathrm{E}({\downarrow} y);$$
$$\vdash \mathrm{E}({\downarrow}(x \vee \sim x));$$
$$\vdash \neg\mathrm{E}({\downarrow}(x \wedge \sim x)).$$

The problem of the existence of classes is solved in a way, different from that for the above cases. For the primary classes: (1) E(Ka) if and only if a is a subject-term; (2) $\neg$E(Ka) if and only if a is not a subject-term; (3) $\sim \mathrm{E}(\mathrm{K}a) \equiv \neg\mathrm{E}(\mathrm{K}a)$. For other class-terms the existence predicate is defined in the following way: E(a) if and only if in constructing a the rules for term construction are observed, and the introduction of a does not lead to contradiction. Let us form, for instance, such as class-term a: 'a class which includes the individual b and does not include the same individual

b'. According to its construction, this term has the following property: $b \in a$ and $\sim (b \in a)$, in other words, it leads to contradiction. By the definition of a itself, such a class does not exist.

As was already said, the situation with the existence of clusters of objects is different. Here everything depends on how the cluster-term is constructed. If a term a is constructed in such a way that the cluster a includes just the objects $b^1, \ldots, b^m$ $(m \geqslant 1)$, then $E(a) \equiv \text{Df.} (Eb^1) \wedge \ldots \wedge E(b^m)$. If a is constructed in such a way that only the number of objects included in a is given, then for the existence of a it is necessary and sufficient that exactly this number of the objects of a given kind exist. If it is not specified what objects of a given kind are included in cluster a and how many objects are included in a, two cases are possible. If in this situation a is an individual term, the following definition holds: E(a) if and only if there exist at least one individual included in a, i.e.,

$$E(a) \dashv\vdash (\exists b)E(b \downarrow (b \in a)),$$
$$E(a) \dashv\vdash (\forall b)\urcorner E(b \downarrow (b \in a)),$$
$$?E(a) \dashv\vdash \sim E(a) \wedge \sim \urcorner E(a).$$

If a is a common (generic) term, the problem is solved in the same way as was done above for the generic terms, i.e., $E(a) \dashv\vdash (\exists b)E(b \downarrow (b \rightharpoonup a))$, etc. (only in this situation the problem for b is solved in the frame of the definitions given above, since b represents a cluster).

In a completely different manner the existence predicate is defined for abstract objects. Two aspects ought to be distinguished here. An object b exists empirically if and only if it has an empirical interpretation $a \rightharpoonup b$, and a does exist. This is one aspect. The other aspect is purely logical: b exists logically if and only if the logical rules are observed in its formation and if its use does not lead to contradiction.

8. Quantifiers and Existence

Consider statements 'Certain objects are called phlogiston' and 'Certain objects are called a round square'. Their structure is $(\exists a)x$. In correspondence with the tradition accepted in logic, these statements are regarded as the following: 'There exists an object which is called phlogiston' and 'There exists an object which is called a round square'; the latter expressions have a structure $E(a \downarrow x)$. But while the first two are considered true (phlogiston is called phlogiston, and round squares are called round squares), the latter are not (one of them is rejected as a physical hypothesis, and the other as logically contradictory). Thus, in the general case, from the statements $(\exists a)x$ the expressions $E(a\downarrow x)$ do not follow logically. Statements $(\exists a)x$ do not imply the existence of a, in other words, the statements $E(a)$ do not follow from them. They imply a weaker requirement, namely, just a possibility of the choice of a. And one also can choose non-existing objects. This can be done, in particular, by using terms, denoting them, the meaning of which is known. From this point of view statements $(\exists a)x$ mean: it is possible to choose such an a that x. Similarly, statements $(\forall a)x$ do not imply the existence of a. They mean: regardless of which a is chosen, x is true with respect to it. Thus, quantifiers $\exists$ cannot be identified with predicates E.

The relation between the quantifier $\exists$ and the predicate E is set up by the rules:

(1) $E(a) \vdash (\exists b)(b \rightharpoonup a)$,
(2) $\neg E(b)(b \rightharpoonup a) \vdash \neg E(a)$,
(3) $E(a\downarrow x) \vdash (\exists a)x$,
(4) $(\neg\exists a)x \vdash \neg E(a\downarrow x)$.

On the other hand, the assertions

$$(\exists b)(b \rightharpoonup a) \vdash E(a),$$
$$(\exists a)x \vdash E(a\downarrow x),$$

do not represent logical rules. If they are accepted in certain situations, they are accepted only as extralogical assumptions. The acceptance of such assertions as general logical rules leads to paradoxes. In addition to what has been said at the beginning of this section, let us discuss another primitive example. From $\neg E(a)$, follows $(\exists a)\neg E(a)$. If $(\exists a)x$ is interpreted as 'There exists a such that x', we obtain 'There exists a such that a does not exist'. And this is equivalent to the assumption which is meant here and according to which from $(\exists a)\neg E(a)$ follows $E(a\downarrow \neg E(a))$. Since $a\downarrow \neg E(a) \leftrightarrow a\downarrow \neg E$, we obtain $E(a\downarrow \neg E)$.

9. Modal Predicates

Predicates 'possible', 'necessary' and 'contingent' (they are called modal) will be abbreviated by symbols M, N and C. As in the case with predicate E, the meaning of these predicates depends on the logical types of objects (subjects) to which they refer.

First of all, it must be noted that the modal predicates are associated with state-terms, i.e., with terms of the form $\downarrow x$. As far as the object-terms are concerned, the use of modal predicates is but a result of the following kind of abbreviations:

$$\mathrm{M}(a) \equiv \mathrm{Df.}\ \mathrm{M}(\downarrow \mathrm{E}(a)),$$
$$\mathrm{N}(a) \equiv \mathrm{Df.}\ \mathrm{N}(\downarrow \mathrm{E}(a)),$$

etc. In what follows, when considering such statements as M(a), N(a) and so on, we shall assume that a represents state-terms.

One can say about statements with modal predicates everything that was said concerning the place, time, conditions, etc. for predicate E. But in order to simplify the presentation, all this, as above, will be omitted as self-evident and implied.

The predicate N is defined through predicate M in the following way:

$$\mathrm{N}(\downarrow x) \equiv \mathrm{Df.}\ \urcorner \mathrm{M}(\downarrow \sim x),$$
$$\urcorner \mathrm{N}(\downarrow x) \equiv \mathrm{Df.}\ \mathrm{M}(\downarrow \sim x),$$
$$?\mathrm{N}(\downarrow x) \equiv \mathrm{Df.}\ ?\mathrm{M}(\downarrow \sim x).$$

And in those cases when the negations $\sim$ and $\urcorner$ coincide, the definition

$$\mathrm{N}(\downarrow x) \equiv \mathrm{Df.} \sim \mathrm{M}(\downarrow \sim x)$$

will suffice.

Thus, in further discussion we shall be concerned only with predicate M.

It must be emphasized that the terms M, N and C are predicates, and not subjects. However, they are frequently considered as subjects in contexts referring to possibility, necessity and randomness as empirical objects. For instance, 'The fact that a has resulted is purely accidental', 'It is possible for B to be included in A', etc. But in such situations either the words 'possible', 'necessary' and 'accidental' are used with the meaning that this has nothing to do with state-modalities (possibilities imply the means, necessity is understood as obligation, etc.) or the expressions containing these words represent stylistic variants of statements containing modal predicates (for instance, the sentence 'There is a possibility for B to be included in A' may represent a modification of the statement 'The fact that B will be included in A is possible').

10. Possibility

In linguistic practice the predicate M is introduced in most varied situations that cannot be unified in one definition. We shall list the most important cases.

For generic terms $\downarrow x$ the predicate M is defined similarly to the predicate E:

$$\mathrm{M}(\downarrow x) \equiv \mathrm{Df.}\ (\exists b)\mathrm{M}(b\downarrow(b \rightharpoondown \downarrow x)),$$
$$\urcorner\mathrm{M}(\downarrow x) \equiv \mathrm{Df.}\ (\forall b)\urcorner\mathrm{M}(b\downarrow(b \rightharpoondown \downarrow x)),$$
$$?\mathrm{M}(\downarrow x) \equiv \mathrm{Df.} \sim \mathrm{M}(\downarrow x) \wedge \sim \urcorner\mathrm{M}(\downarrow x),$$

where b is the variable for states.

For state-terms with operators $\vee$ and $\wedge$ the predicate M is defined by a certain system of axioms, among which the following axioms may occur (the systems of axioms may vary, and those axioms listed below may prove to be theorems deducible from other axioms):

$$\mathrm{M}(\downarrow(x \vee y)) \dashv\vdash \mathrm{M}(\downarrow x) \vee \mathrm{M}(\downarrow y),$$
$$\urcorner\mathrm{M}(\downarrow(x \vee y)) \dashv\vdash \urcorner\mathrm{M}(\downarrow x) \wedge \urcorner\mathrm{M}(\downarrow y),$$
$$\urcorner\mathrm{M}(\downarrow(x \wedge y)) \dashv\vdash \urcorner\mathrm{M}(\downarrow x) \vee \urcorner\mathrm{M}(\downarrow y),$$

etc. It is interesting that in this case explicit definitions are not always possible. Thus, since the assertion

$$\mathrm{M}(\downarrow x) \wedge \mathrm{M}(\downarrow y) \vdash \mathrm{M}(\downarrow(x \wedge y))$$

is unacceptable (and it is unacceptable in the general form, since if $x \to \sim y$, then $\urcorner\mathrm{M}(\downarrow(x \wedge y))$), such a definition as

$$\mathrm{M}(\downarrow(x \wedge y)) \dashv\vdash \mathrm{M}(\downarrow x) \wedge \mathrm{M}(\downarrow y).$$

is also unacceptable.

However, in this case one can find, under certain limitations, implicit definition which can be replaced by the explicit one. In particular, in the above example this can be done in the following way:

$$\mathrm{M}(\downarrow(x \wedge y)) \dashv\vdash \mathrm{M}(\downarrow x) \wedge \mathrm{M}(\downarrow y) \wedge \sim(x \to \sim y).$$

In the case of n-tuples and clusters of objects, the predicate M is defined in a manner similar to that for E (if in the definitions for E the symbol E is replaced everywhere by M, we shall obtain definitions for M). For example,

$$\mathrm{M}(a^1, \ldots, a^n) \equiv \mathrm{Df.}\ \mathrm{M}(a^1) \wedge \ldots \wedge \mathrm{M}(a^n).$$

For abstract objects as well as for classes, the terms E and M are not distinguished, in other words, if a is a term of an abstract object or a class, then

$$\mathrm{M}(a) \equiv \mathrm{Df.}\ \mathrm{E}(a),$$
$$\urcorner\mathrm{M}(a) \equiv \mathrm{Df.} \urcorner\mathrm{E}(a).$$

Finally, consider the meaning of the predicate M in combination with the individual terms $\downarrow x$. One case here is undisputable, namely, the case

$$x \rightarrow \mathrm{M}(\downarrow x).$$

In this case it is implied that x is true, and the predicate M turns out to be practically superfluous. Another case is more interesting, namely when x is not true (or $\sim x$ is true), in other words, $\downarrow x$ does not exist at a time when the statement concerning the possibility or the impossibility of $\downarrow x$ is constructed. But for this case there is no rigorous definition of the predicate M. The skill of operating with it may be developed with individual examples. Below we shall describe a certain general scheme for this.

Let t^1 be a time at which a statement $\alpha\mathrm{M}(\downarrow x)$ is constructed (accepted) (α denotes the presence or the absence of $\urcorner$ or ?). Two subcases are possible: (1) It is assumed that a state $\downarrow x$ is either possible or impossible in the future (at a certain time t^2); (2) it is assumed that a state $\downarrow x$ was either possible or impossible at a certain time t^2 in the past. The second subcase may be reduced to the first one, since the meaning of the predicate M in this situation is established by means of transferring the whole case to a time t^3, preceding t^2. For instance, at t^2 the state $\downarrow x$ did not exist; the predicate M will be considered from the point of view of the situation at a time t^3, preceding t^2, at which $\downarrow x$ also did not exist; but our thinking is such that at t^3 we do not know that $\downarrow x$ will not exist at t^2. And it is not excluded that according to certain rules for introduction of M, the statement $\mathrm{M}(\downarrow x)$, referring to t^2, will be acceptable at t^3. Such instances of usage occur frequently (for example, in the statement 'Napoleon could win the battle of Waterloo').

Thus, what remains is to consider the case of the use of predicate M for individual states that do not exist at the present time; the future time of the existence or non-existence of these states is assumed. The definitions for this case, given below, do not represent the rules for obtaining true (or accepted) statements with the predicate M, but can be regarded only as an explanation of the meaning of this predicate in such a case.

In this situation, the definition should be applied not to the expression $\mathrm{M}(\downarrow x)$, but to the expression $[\mathrm{M}(\downarrow xt^2])t^1]$, in other words, to the expression 'The fact that x at a time t^1'. The definition will take on the following form

$$[\mathrm{M}(\downarrow [xt^2])t^1] \equiv \mathrm{Df.}\ (t^2 > t^1) \wedge \sim[xt^1] \wedge (\urcorner\exists\downarrow a)$$
$$([at^1] \wedge ([at^1] \rightarrow \sim[xt^2])),$$

where a is the variable for statements, $t^2 > t^1$ is an expression 't^2 is later than t^1', $\sim[xt^1]$ means that x is not-true at t^1, and the remaining part of the defining expression means that there is no such statement that would be true at t^1 and from which it would not follow that x is not-true at t^2. Since there is an implied relation between t^1 and t^2, instead of the expression $[M(\downarrow[xt^2])t^1]$, its abbreviated linguistic substitute $M(\downarrow x)$ is used; this leads to the unsolvable problems of its interpretation.

The negation of the possibility of $\downarrow x$ is defined in the following way:

$$[\neg M(\downarrow[xt^2])t^1] \equiv \text{Df. } (t^2 > t^1) \wedge \sim[xt^1] \wedge (\exists\downarrow a)$$
$$([at^1] \wedge ([at^1] \rightarrow \sim[xt^2])).$$

The uncertainty is defined in the standard manner:

$$?M(\downarrow x) \equiv \text{Df. } \sim M(\downarrow x) \wedge \sim\neg M(\downarrow x).$$

11. Contingent Events

The predicate 'contingent' [or 'incidental'] (the symbol 'C' will be used) is defined through the predicate M in the following way:

$$C(\downarrow x) \equiv \text{Df. } x \wedge M(\downarrow \sim x),$$
$$C(\downarrow x) \equiv \text{Df. } x \wedge \neg M(\downarrow \sim x),$$
$$?C(\downarrow x) \equiv \text{Df. } \sim I(\downarrow x) \wedge \sim\neg I(\downarrow x).$$

We must say again that what is meant is the meaning of the predicate C, and not what is responsible for the certainty of the validity of a statement with this predicate. For instance, the statement 'A has lost a chess game contingently to B' may be regarded from two points of view: (1) what is the meaning of the word 'contingent'; (2) why do we believe it. The first question should be answered by logical means; the answer to the second question is outside of the sphere of competence of logic.

One should differentiate between the contingent quality and the modal indifference IND. The latter is defined in the following manner:

$$IND(\downarrow x) \equiv \text{Df. } M(\downarrow x) \wedge M(\downarrow \sim x),$$
$$IND(\downarrow x) \equiv \text{Df. } \neg M(\downarrow x) \vee \neg M(\downarrow \sim x).$$

In many cases of linguistic practice the predicate 'contingent' is used in such a way that statements containing it cannot be verified. This is especially true with respect to individual events ('Napoleon has lost the battle of Waterloo contingently (not contingently)', 'The origin of life on

Earth was contingent (not contingent)', 'The language in human society has appeared contingently (not contingently)', etc.

Consider the following situation. Let a person A be in a certain room x, from which it is possible to go into any of the isolated rooms $y^1,\ldots,y^n$ $(n \geqslant 2)$. After going out in a certain room y^i, A cannot come back to x and go into another room y^k. In the room y^1 there is another person B. Since A can go out of x into y^1, a meeting between A and B can take place. But since A may go out not in y^1, but in a different y^i, it is also possible that the meeting between A and B will not occur. According to the specified conditions and the definition of the contingency predicate, the following is valid: if A has come out of x and entered y^1 and, therefore, his meeting with B has taken place, this was a contingent event.

But let us change the conditions for A. Suppose that there is a possibility for A to go into any of y^i and then come back, and assume that he can do it a given (in particular, unlimited) number of times. And if under these conditions the meeting between A and B did take place, we have no grounds as yet to come up with a statement about this situation containing the contingency predicate.

In this case there is a possibility of variants that give additional information about the situation. In one version the following condition is accepted: the number of A's exits from x into y^i is no less than n, and A has to go into each of y^i. In this case, obviously, A will go into y^1 also, so that its meeting with B will necessarily take place (it is necessary). And if the meeting between A and B is necessary, it is not contingent from the point of view of the definition of the 'contingency' predicate.

The only condition in the second version is that the number of exits, of A from x into y^i is not less than n. Assume that A has done m exits, and has met B at the last one. Was this occurrence contingent or not contingent? This question cannot be answered, since in this case the predicate 'contingent' is not defined. An interesting aspect of this situation is that the cause of the absence of conditions for the use of this predicate is the additional information concerning the case, which excludes the possibility of the acceptance of the statement to the effect that that the meeting between A and B may not take place.

Let a be an event the occurrence of which depends on the occurrences of events from the class b in the following way: if a has occurred, a certain event b has occurred, and if b has not occurred, then a has not occurred. Assume that the class b is divided into n non-intersecting potentially non-empty subclasses, and the cardinality of the class b is equal to m. The predicate C for this case may be defined in the following way: (1) if a has occurred, and if $m < n$, then C(a); (2) if a has occurred and if $m \geqslant n$, then $\sim$ C(a). In this case the expression $\neg$C(a) remains undefined, because

$$\neg C(a) \dashv \vdash N(a),$$

while it does not follow from the conditions specified that a will necessarily occur. Thus, in this situation the distinguishing between negations $\sim$ and $\neg$ is quite essential: $\sim$ C(a) means 'It is impossible to say that a is contingent', and $\neg$C(a) means 'a is not contingent (i.e., a is necessary)'.

Let us return to our example. If the number of exits of A from x is less than the number of different y^i, then, by virtue of the adopted definition, its meeting with B is incidental, and if it equals or is greater than the number of y^i, it would be wrong to say that it is coutingent. But one cannot call it regular. In order to be able to call it regular, additional conditions are necessary, from which the assertion of the necessity of the meeting would follow.

The predicate 'contingent' is also used in cases for which the probability of the occurrence of an event is relatively small. There is no universal limit below the value of which the magnitude of the probability is considered so small that the word 'contingent' is used; this limit depends on specific conditions and the corresponding explicit or implicit assumptions. For instance, the probability of the airplane crash equal to 0.01 in peace time is considered as enormous, while the probability 0.1 in war time as insignificant.

There are still more complex situations when the occurrence of an event is related to a great number of other events, with respect to which one can just make assumptions. For these cases the conditions for the use of the predicate 'contingent' are not established.

12. Fatalism

Fatalism is a conception of the world based on an assumption that all the events in the world occur with necessity. Expanded over the future, it leads to the conception of predetermination.

This conception cannot be regarded as a scientific hypothesis that can be argued against. It results from the ambiguity of predicates M and N.

According to the very definitions of the predicates of possibility and necessity, only the following assertions A are valid:

$$x \vdash M(\downarrow x) \qquad N(\downarrow x) \vdash x,$$

in other words, 'The existing is possible' and 'The necessary does exist or shall exist' (at a certain time), while the assertions B:

$$x \vdash \mathrm{N}(\downarrow x) \qquad \mathrm{M}(\downarrow x) \vdash x$$

are not valid i.e., the assertions 'If something exists, it exists necessarily' and 'If something is possible, it exists' are invalid. It is not because of some experimental study of the world that statements A are valid and statements B are not, but because of the definitions of the terms M and N (i.e. the method of their introduction). Thus, if a certain state $\downarrow x$ does exist (or existed, or shall exist), it does not follow logically from it that $\downarrow x$ is necessary. It is possible that on the basis of some other sources, it will be established that $\downarrow x$ is indeed necessary. Therefore, the statement 'All that existed and exists is necessary' implies a different definition of the predicates of possibility and necessity, according to which

$$\mathrm{E}(a) \vdash \mathrm{N}(\downarrow \mathrm{E}(a)) \qquad \neg \mathrm{E}(a) \vdash \neg \mathrm{M}(\downarrow \mathrm{E}(a)).$$

Also, it would be wrong to believe that everything that has not happened in the past was impossible, since, by virtue of the definition of the predicate M the expression

$$\sim x \vdash \neg \mathrm{M}(\downarrow x)$$

is invalid. On the other hand, the statement $\neg \mathrm{M}(\downarrow x)$ may turn out to be true, but certainty with respect to this fact will be based not on the knowledge that $\sim x$ is true, but on a different piece of information.

As far as future states are concerned, the conception of predetermination either coincides with the tautology 'What will be, will be' or represents a special case of the above definition of predicates.

In order to verify the assertion 'All the events that will take place in the future will necessarily happen (are necessary)' in the case when the above definition of the predicate N is valid, one has to check each statement of the form $\mathrm{N}(\downarrow x)$, where $\downarrow x$ denotes states that are not present now but that will be present in future. But, first of all, we cannot even have terms of the form $\downarrow x$ to denote all future states. Second, in order to accept such statements at present, when $\sim x$ is true, one has to have well justified grounds, and not just purely psychological belief. And these grounds should include the rules for acknowledging the truth of such statements as $\mathrm{N}(\downarrow x)$ for future states. And these rules, in general, reduce to the following statement: if there are such statements y that we accept as true now, and if $y \rightarrow x$ is true, then the statement $\mathrm{N}(\downarrow x)$ is true. If, on the other hand, there are no such statements or if there are statements z such that $z \to \sim x$, then $N(\downarrow x)$ is not-true. But as is well known, although there are cases when statements y exist, these cases are not frequent, and besides, there are always a lot of stipulations, errors, etc. present. Thus, if one accepts a definition of M and N such that assertions A are valid and assertions B are invalid, then the conceptions of fatalism and predetermination appear to be a result of logical errors.

13. Modal Operators

The terms 'possible' and 'necessary' may play the role not only of predicates but also of operators, in analogy with the fact that the terms 'exists' and 'universal' may appear as predicates and quantifiers. We shall use for this purpose symbols M [possible] and N [necessary], respectively. They are defined as follows:

$$(Ma)x \equiv \text{Df. } M(a\downarrow x),$$
$$(\urcorner Ma)x \equiv \text{Df. } \urcorner M(a\downarrow x),$$
$$(Na)x \equiv \text{Df. } N(a\downarrow x),$$
$$(\urcorner Na)x \equiv \text{Df. } \urcorner N(a\downarrow x).$$

14. The Actual, the Existential, the Potential

In the defining part of the definitions of terms, expressions may appear that contain E, M and $\exists$. In this case we shall distinguish between the existential, the potential and the actual presence of the defined attribute. For instance, according to definitions

$$Q^1(a) \equiv \text{Df. } E(c\downarrow P(a,c)),$$
$$Q^2(a) \equiv \text{Df. } M(c\downarrow P(a,c)),$$
$$Q^3(a) \equiv \text{Df. } (\exists c)P(a,c),$$

a is characterized by the attribute Q^1 existentially, by Q^2 potentially, by Q^3 actually. If one accepts $(\exists a)x \to E(a\downarrow x)$, then for objects a the actual coincides with the existential. If $M(a\downarrow x) \to E(a\downarrow x)$ is accepted, then the potential coincides with the existential. If $(\exists a)x \to M(a\downarrow x)$ is accepted, then the actual coincides with the potential. Since in traditional interpretation, the existence was rejected as a predicate and was accepted only as a quantifier, the existential and the actual were not distinguished, and the actual was opposed only to the potential. In our presentation the expression 'actual' is not attached to any other meaning.

15. The Measurement of Possibility

The possibilities of states are differentiated according to magnitude or degree. For instance, such expressions as '$\downarrow x$ is quite possible', 'The probability of $\downarrow x$ is negligible', 'The probability of $\downarrow x$ is close to zero', '$\downarrow x$ is more probable than $\downarrow y$', etc. are used which explicitly express the difference of degrees of possibility. In these cases the predicate M is joined by certain linguistic means α which fix the degree; this results in formation of predicates of the type Mα. These may be, in particular, such predicates as 'M with degree equal to β', 'M with degree greater than β', etc.

A special case of predicates of type Mα is provided by probability predicates, where α is an expression $=\beta, <\beta, >\beta, \geqslant\beta$ or $\leqslant\beta$ and β represent numbers within a certain interval from A to B. Usually probability is measured by numbers ranging from 0 to 1. This is an extralogical fact. We shall consider numbers 0 and 1, respectively, as 'the minimal quantity' and 'the maximal quantity'. Expressions 'The probability of $\downarrow x$ is equal to β (is β, is greater than β, less than β etc.)' just represent linguistic transformations of statements of the form $\mathrm{M}\alpha(\downarrow x)$ ('$\downarrow x$ is possible with a degree β', '$\downarrow x$ has a degree of possibility equal to β', etc.). For the sake of brevity and convenience in reading, we shall write these expressions in conventional symbolic form as $p(x) = \beta$, $p(x) > \beta$, and so on.

The logical properties of the probability predicates are defined by the following statements:

$$\vdash 0 \leqslant p(x) \leqslant 1,$$
$$\vdash (p(x) = 0) \leftrightarrow \neg \mathrm{M}(\downarrow x),$$
$$\vdash (p(x) > 0) \leftrightarrow \mathrm{M}(\downarrow x),$$

which represent a special case of statements for predicates of quantities, and also by the additional statements:

$$\vdash p(x \wedge y) \leqslant \min(p(x), p(y)),$$
$$\vdash p(x \vee y) \geqslant \max(p(x), p(y)),$$
$$\vdash p(\sim x) = 1 - p(x),$$

referring to operators $\wedge$, $\vee$, $\sim$. For the conditionality operator, the following assertions hold:

$$\vdash (x \rightarrow y) \rightarrow (p(y) \leqslant p(x)),$$
$$\vdash x \rightarrow (p(x \rightarrow y) = p(y)).$$

The existence of various methods for computation of probability, and the possibility of modifying the computation rules, cannot be considered as ground for different notions of probability. The point is that the word 'probability' is just an abbreviation for the expression 'the magnitude (or the degree) of possibility' while its exact definition should not be based on such formulations as 'the probability is...', but on the description of the properties of the probability predicates and, moreover, of statements containing these. The correct formulation of the question is as follows: what is the meaning of the expressions $M\alpha(\downarrow x)$, $p(x)=\alpha$, $p(x)>\alpha$, etc.? The logically correct answer to this question consists in description of the properties of the predicates M, and in description of ways of determining α.

There is an expression in common use 'the probability of a statement'. It has multiple meanings and uncertainty. It may mean the probability of the fact that a certain statement x is either true or will be true, and these expressions, in turn may mean that $\downarrow x$ either exists or will take place, and also that we are right to make a statement x with respect to a certain object.

Two cases are of interest here. The first one: the probability that $\downarrow x$ exists is the probability that we shall detect the existence of $\downarrow x$. Here we have in mind already not the state $\downarrow x$, but the state $\downarrow y$, where y is an expression 'The existence of x is detected (established, etc.)'. The following obvious rule holds for this case: $p(y) \leqslant p(x)$.

The second case: the probability that we are right to make a statement x. This is a special case of calculating the probability for individual states. In this situation one takes into account both the circumstances in favor of x and those against it. Only this case is more complex because the characteristics of the objects mentioned in x may appear as the arguments 'for' and 'against' (for instance, when we establish the probability that a passer-by is A, we take into consideration the known characteristics of A and his life circumstances).

A few more words about an interesting logical error. Consider a statement to the effect that a possible event will sooner or later take place. Indeed, if $M(\downarrow x)$, then $p(x)>0$. And if the number of specific 'trials' in which the state $\downarrow x$ may occur is sufficiently great, the event $\downarrow x$ will sooner or later take place. But this would have been true only if the frequency method were the only way to establish the value of the probability of events. As we have seen, this method is not applicable to individual events, and, in general, to those cases when the number of trials cannot be sufficiently great. Besides, the statement $p(x) > 0$ may simply be a consequence of $M(\downarrow x)$, and the latter may be accepted only because x does not lead to a contradiction, because there are no such $\downarrow y$ that $y \rightarrow \sim x$, and so on.

16. Relations

There are statements among those with two or more local predicates, which have the structure $a\mathrm{R}b$ and $a \urcorner \mathrm{R} b$, or may be reduced to this form through certain linguistic transformations. For instance, '*a* is greater than *b*', '*a* is not greater than *b*', '*a* is three times heavier than *b*', etc. Such statements are called relation statements, and their subject-matter is called relations (i.e., if x is a relation statement, the object, denoted by the term $\downarrow x$, represents the relation). For example, the term 'The fact that *a* is greater than *b*' denotes a relation.

In symbolic expressions $a\mathrm{R}b$ and $a \urcorner \mathrm{R} b$, the symbol R denotes the type of relation predicate (and. correspondingly, of relation), but not the predicate in its complete form. Thus, in the statement '*a* is greater than *b*' the subjects are the terms *a* and *b*, the predicate represents the expression 'the first object is greater than the second', the type of relation is denoted by the term 'greater'. The term *b* may coincide with a pair, a triplet, etc. of terms. For instance, in the statement '*a* is between *b*, *c* and *d*' the term, following the relation 'is between', is a triplet of terms (b, c, d).

In the case of relational statements, the operator ? is defined in the following way:

$$(a ? \mathrm{R} b) \equiv \mathrm{Df.} \sim (a\mathrm{R}b) \wedge \sim (a \urcorner \mathrm{R} b).$$

A relation R is reflexive if and only if $a\mathrm{R}a$ is true; symmetric if and only if $(a\mathrm{R}b) \rightarrow (b\mathrm{R}a)$ is true; and transitive if and only if $(a\mathrm{R}b) \wedge (b\mathrm{R}c) \rightarrow (a\mathrm{R}c)$.

17. Comparison Relations

Among the relations, we differentiate between comparison relations (for instance, 'heavier') and ordering relations (for instance, 'is located to the right of'). In the case of comparison, certain attributes of objects are identified, the presence or the absence of certain characteristics of objects (in other words, similarity and difference between them), as well as qualitative similarity or difference with respect to a certain attribute, are established.

There is a class of comparison statements referring to the superiority of one kind of objects with respect to another in terms of a certain characteristic. The symbolic form of these statements will be

$$a > pb,$$

where the symbol $> p$ reads 'is superior with respect to the attribute p'. This is a schematic or generalized representation of a group of statements, the stylistic forms of which can be most diversified and for which relation symbols are not always divided into parts $>$ and p. Thus, in the statements 'a is higher than b' and 'a is heavier than b' the words 'heavier' and 'higher' represent special cases of the relational terms 'is superior in height' and 'is superior in weight'. The internal negation and the uncertainty form of such statements are

$$a\urcorner > pb \text{ and } a? > pb.$$

The relations 'is identical with respect to an attribute' and 'is inferior with respect to an attribute' are defined through relation of superiority. We shall write them symbolically as $=p$ and $<p$. As examples: the statements 'a has the same height as b', 'a is equal in weight to b', 'a is lighter than b', 'a is smaller than b', etc. These relations may be explicitly defined as

$$(a = pb) \equiv \text{Df. } (a\urcorner > pb) \wedge (b\urcorner > pa),$$
$$(a\urcorner = pb) \equiv \text{Df. } (a > pb) : (b > pa),$$
$$(a < pb) \equiv \text{Df. } (b > pa),$$
$$(a\urcorner < pb) \equiv \text{Df. } (b > pa) : (a = pb),$$

where the operator ':' is that of strong disjunction ('one and only one of the two') Uncertainty is defined according to the general rule.

The logical properties of the relation symbol $> p$ are defined implicitly by a system of axioms, including the following:

(1) $\vdash (a\urcorner > pa),$

(2) $(a > pb) \vdash (b\urcorner > pa),$

(3) $(a\urcorner > pb) \vdash (b > pa) : (a = pb),$

(4) $(a > pb) \wedge (b > pc) \vdash (a > pc),$

(5) $(a > pb) \wedge (b = pc) \vdash (a > pc),$

(6) $(a = pb) \wedge (b > pc) \vdash (a > pc),$

(7) $(a\urcorner > pb) \wedge (b\urcorner > pc) \vdash (a\urcorner > pc).$

The first axiom means that the relation $>p$ is non-reflexive, the second that it is non-symmetric, the fourth that it is transitive. The symbolic expression $A = pB$ appears here only as abbreviation for $(A \neg > pB) \wedge (B \neg > pA)$, and it may be eliminated.

The following circumstance is extremely important. Comparison of objects is always performed by a certain method and with respect to a third something which is different from the compared objects. For instance, in order to be able to say that *a* is moving more slowly than *b* (or that the velocity of *a* is less than the velocity of *b*), one has either to possess a certain method of comparison of quantities or to have a method of observing the moving objects, in other words, a person who performs the comparison, as well as means for this operation, is necessary. Thus, we shall agree to consider the letter following the symbols $>$, $=$ and $<$ not only as a symbol of the objects' attribute but also as a symbol of the method of comparison, i.e., the method of comparison in our presentation will include both the attributes with respect to which the comparison is made and the way of using these attributes in comparison

A few words about the evaluation expressions, since this issue is related to that of comparison. Examples of evaluation expressions are 'good', 'bad', 'better', 'worse', 'progressive', and so on, and also evaluation points (like examination marks, and the points in certain sport contest events). Among evaluations, one differentiates between absolute and relative evaluations. Both these groups imply certain evaluation criteria. Absolute evaluations are constructed according to the following scheme: (1) there are certain evaluation criteria V; (2) the comparison of an individual object *a* with V results in P(*a*), where P represents the absolute evaluation of *a* in comparison with V. Relative evaluations are divided into two groups. Direct relative evaluations are constructed according to the following scheme: (1) there are evaluation criteria W; (2) the comparison of an *n*-tuple of objects (in particular, of a pair) $a^1, \ldots, a^n$ (where $n \geq 2$) with W results in $P(a^1, \ldots, a^n)$, where P is the direct relative evaluation of objects $a^1, \ldots, a^n$. The scheme for construction of comparative relative evaluations is the following: absolute evaluation of $a^1, \ldots, a^n$ in relation to W results, respectively, in $P^1(a^1), \ldots, P^n(a^n)$; (2) comparison of $a^1, \ldots, a^n$ with respect to $P^1, \ldots, P^n$ results in $R(a^1, \ldots, a^n)$. Examples: (1) a student N has received a mark of 5 in an exam, corresponding to the adopted evaluation criteria V, an absolute evaluation; (2) chess players N^1 and N^2 frequently play chess with each other, and N^1 wins more frequently than N^2; it is said that N^1 is the stronger chess player, a direct relative evaluation; (3) gymnasts N^1 and N^2 independently received evaluation points equal to 9.7 and 9.1, respectively; it is said that N^1 as a gymnast is better than N^2, a relative comparative evaluation.

18. Order Relations

We shall consider the order relation in somewhat greater detail, since it plays an important role in further discussion.

The expression 'order of objects' will be accepted as intuitively clear, based only on examples and explanations. In particular, the location of objects in space as well as their appearance or disappearance in time represent particular cases of order. Such expressions as 'first', 'second',..., 'higher than', 'lower than', 'before', 'simultaneously with', 'to the right of', etc. are used to fix order.

The order of objects *a* and *b* is defined with respect to a third object *c*, which we shall call the reference point, and the point of definition, of the order.

When one establishes the order of objects, not only the reference points are important, but also the methods of particular actions. We may, for instance, define the order of points *a* and *b* on a circle with respect to a third point *c*. But the order of *a* and *b* is not yet specified by this: one still has to indicate the direction of motion (clockwise or counterclockwise). Depending on the direction of motion, different results are obtained: in one case it turns out that *a* will be further than *b* with respect to *c*, and in another, *b* will be further than *a*. In what follows we shall use the expression 'the method of determining order' assuming that the method of determining order also includes the determination of the reference point.

The symbolic form of statements to the effect that *a* is in relation of order R to *b* with respect to the method of determining order α is

$$a(\mathrm{R}\alpha)b.$$

Expressions $a\urcorner(\mathrm{R}\alpha)b$ and $a?(\mathrm{R}\alpha)b$, respectively, represent its partial negation and the uncertainty form.

There are different aspects to the question of the dependence of the order relation between objects on the method of determining order.

The logical aspect: Let α and β be different methods of establishing order. In this situation from $a(\mathrm{R}^1\alpha)b$ logically follows neither $a(\mathrm{R}^1\beta)b$ nor its negation $\sim(a(\mathrm{R}^1\beta)b)$. In these expressions the time t^1 is assumed to be the same. If at this moment of time both $a(R^1\alpha)b$ and $a(R^2\alpha)b$ hold, then, choosing instead of α a different method β, we cannot change the relation $a(R^1\alpha)b$, in other words, to make $\sim(a(R^1\alpha)b)$ true. The reason for this is the rule of logic $\vdash \sim(x \wedge \sim x)$.

The empirical (or physical) aspect: Assume that $a(R^1\alpha)b$ holds at a time t^1 and $a(R^2\beta)b$, at a time t^2. It is possible that for some reason the order relation between a and b at a time t^2 will differ from that at a time t^1, so that $a(R^1\alpha)b$ will not be true. This may happen, in particular, because of the application of method β. But it is not necessarily so in every case: at a time t^2 it is possible that $\sim(a(R^1\alpha)b)$, although is not applied here at all. A situation is also possible for which application of β excludes application of α. And then at the moment t^2 the expression $a?(R^1\alpha)b$, and, therefore, the expression $\sim(a(R^1\alpha)b)$ will be true, whereas $a\urcorner(R^1\alpha)b$ will not.

In some cases it is not difficult to establish the order relations for objects. Their order is specified in a certain manner, and no argument arises. In other cases, order is established through application of the rules of logic, mathematics and the special rules formulated in various scientific fields in connection with the peculiar features of the objects under study (such as the theory of relativity in physics).

As is true with comparison, there are statements in use referring to the superiority in an order of one kind of object over another with respect to a certain method of establishing order, and also to the identity in order of objects and to the fact that some objects are inferior in order to others (again with respect to a certain method of establishing order). The symbolic form of these statements is as follows:

$$a > \alpha b, \quad a < \alpha b, \quad a = \alpha b.$$

For their internal negations and uncertainty form we have:

$$a\urcorner > \alpha b, \quad a? > \alpha b, \quad a\urcorner < \alpha b,$$

etc. The difference between this case and that of comparison is just that the method of establishing order is substituted in place of the attribute p.

Relations $=\alpha$ and $<\alpha$ are defined through $>\alpha$ in exactly the same manner as in the case with similar relations of comparison, while the relation $>\alpha$ is defined implicitly through axioms of order; the latter include those analogous to the comparison axioms (1)–(7) for $>p$ (the only difference being that the ordering method α is substituted in place of p).

Note that distinguishing between $>\alpha$ and $<\alpha$ does not predetermine the solution of the problem which of the pair relations 'further – closer', 'higher – lower', 'to the right – to the left', etc. will be $>\alpha$ and which will be $<\alpha$. The only circumstance of importance here is that if one of them is regarded as $>\alpha$, the other one is $<\alpha$.

19. The Relation 'Between'

The definition of the simplest case of the relation 'between' is as follows: a is between b and c with respect to α if and only if $(a > \alpha b) \wedge (c > \alpha a)$ or $(b > \alpha a) \wedge (a > \alpha c)$. In its general form (for an arbitrary number of objects) this relation is defined as: a is between $b^1,\ldots,b^n$ with respect to the methods of establishing order $\alpha^1,\ldots,\alpha^m$ if and only if for any pair b^i and b^k among $b^1,\ldots,b^n$ there will be α^j among $\alpha^1,\ldots,\alpha^m$ such that $(a > \alpha^j b^i) \wedge (b^k > \alpha^j a)$ or $(b^i > \alpha^j a) \wedge (a > \alpha^j b^k)$.

The following is valid for clusters (and n-tuples) of objects: a cluster (an n-tuple) of objects A is between $b^1,\ldots,b^n$ with respect to $\alpha^1,\ldots,\alpha^m$ if and only if every object included in A is between $b^1,\ldots,b^n$ with respect to $\alpha^1,\ldots,\alpha^m$.

The symbolic form of the expression 'a is between $b^1,\ldots,b^n$ with respect to α' will be

$$\sqcup(a(b^1,\ldots,b^n)\alpha).$$

20. The Existence of Relations

The question of the existence of relations deserves special attention, since it lies at the heart of the question of the existence of space and time.

The general theses concerning the existence of states $\downarrow x$ are valid in the case of relations. The peculiar features of these states, however, are not taken into account here. Therefore some additional explanations are needed.

A term $\downarrow(a\mathrm{R}b)$, as is obvious from the above, is individual if and only if a and b are individual terms. If $\downarrow(a\mathrm{R}b)$ is an individual term, then the existence predicate for it has the following meaning: the relation $\downarrow(a\mathrm{R}b)$ exists if and only if both a and b exist, and the result of certain operations performed over a and b (comparison of a and b or establishing their order) is exactly R. In the case of general terms $\downarrow(a\mathrm{R}b)$ the problem of existence is reduced to that for individual terms by virtue of the rules for arbitrary general terms.

It follows from the definition of the existence of relation $a\mathrm{R}b$, that if this relation exists, both objects a and b do exist, and if at least one of a

and b does not exist, their relation does not exist either. In other words,

$$E(\downarrow(aRb)) \dashv\vdash E(a) \wedge E(b) \wedge (aRb),$$
$$\neg E(\downarrow(aRb)) \dashv\vdash \neg E(a) \vee \neg E(b) \vee (a \neg R b).$$

21. Ordered Series

A cluster of individuals, such that for any pair a and b among them $a > \alpha b$ or $b > \alpha a$, will be called a series ordered with respect to α. The individuals, included in a given ordered series, represent its elements. As is obvious from the definition, if $a = \alpha b$, then a and b cannot represent the elements of a single series. They can belong only to different series.

Figure 1 depicts three ordered series A, B and C. All of them are ordered with respect to the individual x, and the arrow indicates the direction of selection of individuals and the increase in their order. In Figure 2 we present the ordered series D, E and F, the elements of which are ordered with respect to x, but by a somewhat different method: we determine the

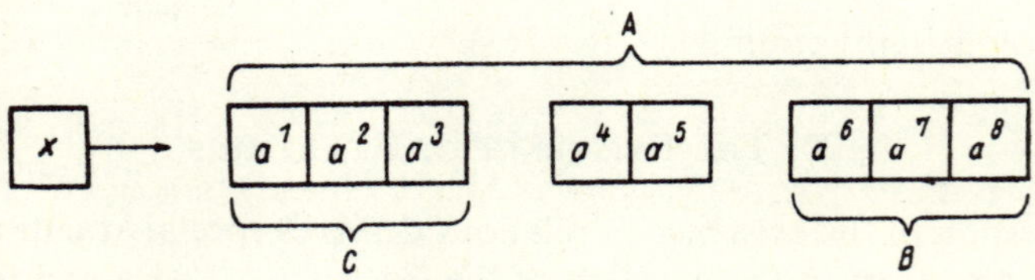

Fig. 1.

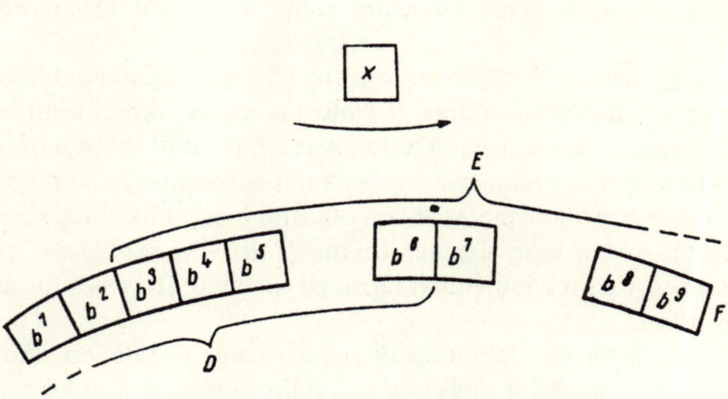

Fig. 2.

rotation of x, with the curved arrow indicating the direction of increase of order of individuals.

Elements a and b of the series A, ordered with respect to α, will be called adjacent if and only if there is no element in A which is between a and b with respect to α. In Figure 1 the adjacent pairs of elements are a^1 and a^2, a^2 and a^3, a^4 and a^5, etc.

The adjacent elements of a series can also be defined in the following way: a and b are the adjacent elements of A if and only if

$$(\neg\exists c)((a \in \mathrm{A}) \wedge (b \in \mathrm{A}) \wedge \sqcup(c(a,b)\alpha,$$

where a, b and c are individual variables. This is the definition of the actually adjacent elements of a series. If in this definition one replaces $\exists$ by M, one obtains the definition of the potentially adjacent elements of a series. In what follows we shall give only the formulations of the definitions with $\exists$ assuming that the reader can make the addition of the word 'actual' and construct the corresponding definition with M.

The fact that a and b represent the adjacent elements of a series with respect to α will be written in the symbolic form as

$$a \mid \alpha b.$$

The relation between series A and B in Figure 1 is denoted by the expression 'a segment of a series'. Here B represents a segment of the series A. Similarly, series E and D in Figure 2 are segments of the series F. The general definition of the term 'a segment of a series' is as follows: let a and b be the elements of a series A; let B be a cluster of individuals such that all individuals of B are not inferior in order to a, with respect to α or not superior in order to b, with respect to α (or both). If $\mathrm{B} \subset \mathrm{A}$, then B is a segment of the series A.

Among the elements of the series A one can choose such elements which form a new series B. For instance, elements a^1, a^3, a^5 and a^7 of the series A in Figure 1 form a new series. But this series cannot be called a segment of A. This is a different relation between series.

A series represents a cluster of individuals, and its existence depends on the way of defining this cluster. For example, a series consisting of individuals a, b and c, ordered in such a way that $a > \alpha b$ and $b > \alpha c$ exists if and only if all three individuals a, b and c exist and are ordered in exactly the same way as described above. But assume that the series A is defined in the following manner: if $a > \alpha b$, then $a \in \mathrm{A}$ and $b \in \mathrm{A}$. For the existence of such a series it is necessary and sufficient that at least one pair of individuals a and b such that $a > \alpha b$ exists.

22. Contact

We shall say that individuals a and b are in contact (are not in contact) with respect to α and a class of individuals A if and only if $a > \alpha b$ or $b > \alpha a$ and no individual (a particular individual) of the class A is possibly to be placed between a and b with respect to α. In Figure 1 individuals a^1 and b are in contact with respect to α and individuals a^3 and a^4 are not in contact since a particular individual b can be placed between them.

Certainly, the question of whether or not one can, in a concrete case, place an individual c between a and b is answered depending on some agreement, practical possibilities, and so on. For instance, Figure 3 depicts two individuals a and b which can be regarded as being in contact on the basis that they have a point of contact k, and at the same time can be regarded as not being in contact since it is possible to place an individual c as is shown on the picture. But this problem is solved practically in every case, and this is sufficient to justify our definition. From the point of view of the definition it is indifferent whether a certain operation can or cannot be realized, because the purpose of the definition is to establish the meaning of terms rather than the rule for determining the truth values of statements containing these terms.

We shall omit references to class A assuming that in each case a certain or an arbitrary class of individuals is meant. A statement 'a is in contact with b with respect to α' will be symbolically written as

$$a \parallel \alpha b.$$

The above definition can be expressed as follows:

$$(1) \qquad (a \parallel \alpha b) \dashv\vdash ((a > \alpha b) \vee (b > \alpha a)) \wedge (\urcorner \mathrm{M}c) \sqcup (c(a, b)\alpha),$$

$$(2) \qquad (a \urcorner \parallel \alpha b) \dashv\vdash ((a > \alpha b) \vee (b > \alpha a)) \wedge (\mathrm{M}c) \sqcup (c(a, b)\alpha),$$

$$(3) \qquad (a ? \parallel \alpha b) \dashv\vdash \sim (a \parallel \alpha b) \wedge \sim (a \urcorner \parallel \alpha b).$$

If operator M is not introduced, in these definitions one should replace M with $\exists$. The term I may be replaced with an individual variable.

In our case it is extremely important to distinguish between negations

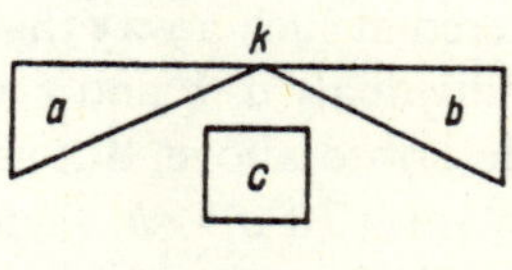

Fig. 3.

$\urcorner$ and $\sim$, since $a \urcorner \| \alpha b$ means 'a and b are not in contact with respect to α', and $\sim(a \| b)$ means 'it is impossible to say (accept) that $a \| \alpha b$' (it may be not only $a \urcorner \| \alpha b$, but also $a = \alpha b, a ? > \alpha b$, etc.).

As follows from the definition,

$$(a = \alpha b) \vdash \sim (a \| \alpha b),$$
$$\sim (a \| \alpha b) \dashv\vdash \sim (a > \alpha b) \wedge \sim (b > \alpha a) \vee,$$
$$\sim (\urcorner \mathrm{MI})(((a > \alpha \mathrm{I}) \wedge (\mathrm{I} > \alpha b)) \vee ((b > \alpha \mathrm{I}) \wedge (\mathrm{I} > \alpha a))),$$
$$(a \| \alpha b) \vdash (a | \alpha b).$$

23. Continuity and Discontinuity of an Empirical Series

A series is called empirical if and only if it is formed by empirical individuals. A series is called abstract if and only if it is formed by abstract individuals.

The same predicates may be defined differently for abstract and empirical series. This is true, in particular, for the predicates of continuity and discontinuity.

A series A will be called continuous with respect to α if and only if all its adjacent elements are pairwise in contact with respect to α. In other words, a series A is continuous with respect to α if and only if

$$(\forall a)(\forall b)((a \in \mathrm{A}) \wedge (b \in \mathrm{A}) \wedge (a | \alpha b) \rightarrow (a \| \alpha b)),$$

where a and b are individual variables. A series A will be called discontinuous (porous) with respect to α if and only if it is true for at least one of its pairs of adjacent elements that they are not in contact with respect to α. To put it differently, a series is discontinuous with respect to α if and only if

$$(\exists a)(\exists b)((a \in \mathrm{A}) \wedge (b \in \mathrm{A}) \wedge (a | \alpha b) \wedge (a \urcorner \| \alpha b)).$$

For instance, series B and C in Figure 1 are continuous, while series D and E in Figure 2 are discontinuous.

24. The Beginning and End of a Series

A series A has a first element with respect to α if and only if there is such an element a of this series that any other element of the series is superior to it in order with respect to α. If a and b are individual variables, the specifying part of this definition is written as

$$(\exists a)(\forall b)((a \in \mathrm{A}) \wedge ((b \in \mathrm{A}) \wedge \sim (a \rightleftharpoons b) \rightarrow (b > \alpha a))).$$

The element a is the starting element, or the first element in order of the series A.

A series A has a final element with respect to α if and only if

$$(\exists a)(\forall b)((a \in \mathrm{A}) \wedge ((b \in \mathrm{A}) \wedge \sim(a \rightleftarrows b) \rightarrow (a > \alpha b))),$$

(in other words, if there is such an element of this series a which is superior to all others in order with respect to α.) The element a is the final element or the last element of the series A with respect to α.

A series A has no starting element with respect to α if and only if

$$(\forall a)(\exists b)((a \in \mathrm{A}) \rightarrow (b \in \mathrm{A}) \wedge (b < \alpha a)).$$

A series A has no last element with respect to α if and only if

$$(\forall a)(\exists b)((a \in \mathrm{A}) \rightarrow (b \in \mathrm{A}) \wedge (a < \alpha b)).$$

A series may be defined in such a way that it is specified as to which element is the starting one, and which one is the last, or both. For example, in Figure 1 the individual a^1 is the starting one for the series A while the individual a^8 is its last one; the individual b^3 is the starting one for series E, and the individual b^6 is the final one for series D. But there may be a definition of series such that it is not specified which exactly of their elements are the final and starting ones. For instance, in the case of the series E in Figure 2 it is not indicated which element is the final, and for the series D, which one is the starting. Sometimes a certain research is carried out, resulting in the establishment of the starting or the final element of a series (or both). In other cases one determines or assumes the impossibility of doing this or in general assumes the absence of the starting or the final element of a series (or of both).

There is a chance that, after a certain moment, it becomes impossible to establish the order of individuals of the series A with respect to α. This corresponds to the case of uncertainty. If x represents the statement 'A series A has the starting (final) element', and y represents the statement 'A series A has no starting (final) element', then the above case will be written as $\sim x \wedge \sim y$. In practice, the situation here is based on one or another kind of agreement. In particular, the final element of the series A, for which the order with respect to α can be established, may be interpreted as the first or the last element of the series.

If, on the other hand, we assume that a series has no starting element or that it has no final one, and it is not clear from α whether or not there will be a moment when it would be impossible to establish the order of individuals with respect to α then it appears that there are series having no starting or no final elements (or even both). However, this is meaningful only at the level of assumptions. As far as such statements as 'a series A has no starting element', 'a series A has no final element', etc. are concerned, they cannot be empirically verified in the cases under consideration, due to the impossibility of considering, element by element, the infinite series of objects and of establishing the relation of order for them. Thus, if one cannot deduce from the definition of a series or from the method of its construction the presence of the starting or the final element, and it is impossible to verify empirically the corresponding statements, then the assumptions of the absence of the starting or the final element, as well as the assumptions of their presence or possibility, are equally justified. But one must remember that those are just assumptions, and not indisputable truths.

The fact of the impossibility of establishing the starting (final) element of a series does not always coincide with their absence. This depends on the definition of the series.

The final element of one series is not always the starting element of another, while the starting element of one series is not always the final element of another. Thus, there may be a situation with a series A with respect to α such that from a certain moment it becomes impossible to establish the order of individuals with respect to α, so that one cannot talk about any starting or final element of the new series with respect to α. As far as a series B, constructed with respect to another way of ordering β, is concerned, from the statement of the existence of the starting or the final element of a series A one cannot deduce the statement of the existence of the final or the starting element of the series B. Thus, a statement to the effect that the final (starting) element of one series represents the starting (final) element of another is either meaningless or wrong.

If a series has no starting element, this does not mean that this series and its elements do not exist.

We shall say that a series is limited if and only if it has both starting and final elements, and that it is unlimited if and only if it has no starting or final elements (or both). As follows from the definitions, it is possible to conceive of a series with respect to which one would be wrong to say that it is limited and it would be wrong to say that it is unlimited (the case of uncertainty).

25. Interval

It is impossible to construct for the expressions 'an interval' and 'an interval between a and b' a definition according to the principle 'An interval (an interval between a and b) is called...'. The only possible thing here is to formulate definitions of complex expressions containing the word 'interval' and statements with this word. For example, an expression 'The length of the interval between a and b is equal to ten c' is a substitute (by definition) for the expression 'It is possible to place ten objects of class c between a and b in such a way that the objects a, b and c form a continuous series'; the expression 'a has occurred in the interval between b and c' is a substitute for the expression 'a has occurred between b and c' (the word 'interval' here is superfluous).

It is appropriate to use the expression 'interval' only in those cases when the objects represent elements of an ordered series. To put it more precisely, the word 'interval' is just a part of the expression 'An interval between a and b with respect to α', which is appropriate exclusively in those situations when $a > \alpha b$ or $b > \alpha a$. For the sake of brevity and visual obviousness, we shall write this expression in symbolic form as

$$\{a, b, \alpha\}.$$

The objects a and b will be called the limits of the interval. The expression $\{a, b, \alpha\}$ represents a subject-term. The terms $\{a, b, \alpha\}$ and $\{b, a, \alpha\}$ are not always identical in meaning; similarly for terms $\{a, b, \alpha\}$ and $\{a, b, \beta\}$, if α and β are different.

A term $\{a, b, \alpha\}$ will be regarded as individual for a given α if and only if a and b represent individual terms. In the final analysis, all other terms of intervals are defined through terms of the form $\{a, b, \alpha\}$, according to the general rules for introduction of terms.

26. Length

Intervals are characterized by their magnitude or length. Such expressions as 'Length of $\{a, b, \alpha\}$' will have the following symbolic form:

$$l\{a, b, \alpha\},$$

where l is just an abbreviation for the word 'length'.

The length of the interval between empirical objects is determined by the possibility or impossibility of placing certain objects between them, and by the number of these objects, and also by the results of comparison with another interval. In this case the method of determining the length of the interval is at the same time a method of determining the expression referring to this quantity, a typical example of 'operational' definitions. For instance, in the expression 'The interval between *a* and *b* equals ten *c*' the length of the interval, 'ten *c*', is specified, while in the expression, through which it is defined ('between *a* and *b* one can place...'), a certain method is given for the measurement of the length of the interval.

The comparison of the intervals $\{a,b,\alpha\}$ and $\{c,d,\beta\}$ implies a certain method of their mutual arrangement such that *a*, *b*, *c* and *d* may be imagined as ordered with respect to α and β, and either $a=\alpha c$ (either $a=\beta c$) or $b=\alpha d$ (or $b=\beta d$), but the intervals should not change their length, in other words, the comparison implies the superposition of one interval over another when one pair of their elements is matched. Let us agree to superpose $\{c,d,\beta\}$ over $\{a,b,\alpha\}$ in such a way that the length of $\{c,d,\beta\}$ is equal to that of $\{c,d,\alpha\}$.

The following implicit definitions of expressions, comparing intervals, are possible:

$$(l\{a,b,\alpha\}=l\{c,d,\alpha\})\dashv\vdash((a=\alpha c)\leftrightarrow(b=\alpha d)),$$
$$(l\{a,b,\alpha\}>l\{c,d,\alpha\})\dashv\vdash((a=\alpha c)\rightarrow(b>(b>\alpha d)),$$
$$((b=\alpha d)\rightarrow(c>\alpha a)),$$
$$(l\{a,b,\alpha\}<l\{c,d,\alpha\})\dashv\vdash(l\{c,d,\alpha\}>l\{a,b,\alpha\}).$$

If α and β are different, the intervals $\{a,b,\alpha\}$ and $\{a,b,\beta\}$ are not necessarily equal.

To denote the dimensions of individual concrete objects, such expressions as 'length', 'height', 'duration', etc. are used. The expression 'length of *a*' will be written briefly in symbolic form as

$$la.$$

The dimensions of empirical objects are defined (where appropriate) in the following manner: the dimension of an empirical object *a* with respect to a certain method of establishing order α is equal to the length of the interval $\{b,c,\alpha\}$ such that if *a* is placed between *b* and *c*, then *b* and *a* will be in contact with respect to α, and also *a* and *c* will be in contact with respect to α. A special case occurs when one takes as *b* and *c* the parts of *a* such that all other parts of *a* are between *b* and *c* with respect to α.

The dimensions of empirical objects are established with respect to certain methods of determining order. If *a* represents an object instead of an interval between objects, then the definition should contain the expression '*la* with respect to α'. It will be written in symbolic form as

$$l\{a, \alpha\}.$$

For different α and β it is not always that $l\{a,\alpha\} = l\{a,\beta\}$.

We can write the above definition of $l\{a,\alpha\}$ in the following form:

$$\vdash (b \parallel \alpha a) \wedge (a \parallel \alpha c) \rightarrow (l\{a,\alpha\} = l\{b,c,\alpha\}).$$

Or as

$$\vdash (l\{a,\alpha\} = l\{b{\downarrow}x, c{\downarrow}x, \alpha\}),$$

where x is $(b \parallel \alpha a) \wedge (a \parallel \alpha c)$.

In the case of abstract objects and the intervals between them, statements concerning dimensions are accepted as assumptions. The only limiting factors here are the logical consistency of statements and aspects of applications of theoretical constructions. The cases are known when objects are permitted that have no dimensions, infinitesimal and infinite intervals, zero length intervals, etc. For empirical objects such assumptions are unjustified. For example, it is impossible to observe objects having no dimensions or having infinitesimal dimensions; it is impossible to record the infinite interval between empirical objects, and so on. All such assumptions can neither be confirmed nor disproved experimentally.

If there is no method of establishing (determining, measuring) the dimensions of certain empirical objects and intervals between them, the corresponding expressions have no meaning. We would like to emphasize the fact that in such cases the meaning of expressions *la* is not established (or even cannot be established in principle). But this is by no means equivalent to the fact that these objects and intervals have no dimensions, in other words, that $lx = 0$.

Note that the expressions of the form $la > lb$, $la < lb$, etc. are identical, respectively, to such expressions as $a > lb$, $a < lb$, etc. (i.e., '*a* is superior to *b* in length', etc.).

If $\sim (a > \alpha b) \wedge \sim (b > \alpha a)$, the concept of the interval cannot be applied in this case. In this situation the answer to the question, what is the interval between a and b, is just the following: $\sim (a > \alpha b) \wedge \sim (b > \alpha a)$. And if it is absolutely necessary to establish the interval between a and b, one has to find such a method β of ordering a and b, according to which $((a > \beta b) \vee (b > \beta a))$.

27. Abstract Series

An abstract ordered series is a series the elements of which are abstract objects.

If the elements of an abstract series have no dimensions, many predicates for them are defined in a manner different from that in the case of the empirical series. This is of paramount importance for elimination of a whole number of difficulties (like Zeno's paradoxes). The latter become absolutely insurmountable if one does not take into account the difference between abstract and empirical series, and the fact that the same predicates are defined differently depending on which series they belong to.

Let a, b and c be the variables for terms, denoting elements of a series A, i.e., $a \in \mathrm{A}$, $b \in \mathrm{A}$, $c \in \mathrm{A}$.

Adopt the following definitions: (1) a series A is continuous with respect to α if and only if

$$(\forall a)(\forall b)(\exists c) \sqcup (c(a, b)\alpha);$$

(2) a series A is discontinuous with respect to α if and only if

$$(\exists a)(\exists b)(\neg \exists c) \sqcup (c(a, b)\alpha).$$

It is obvious that these definitions differ from the corresponding definitions for empirical series. Let us consider the consequences of such an interpretation of an abstract series in terms of an empirical series, which is characterized by the following: for each element a of the abstract series an element b of the empirical series is found such that $b \rightharpoonup a$ is true (in other words, each element of the abstract series is set in correspondence with an element of the empirical series). In this case the continuity of the empirical series will manifest itself in the fact that apparently, for any two elements of the empirical series there is such an element that can be placed between them. And this will go on indefinitely. But this is wrong.

The reason for the error is the incorrect interpretation of the abstract series: it is not only its elements but also its predicates, introduced for abstract objects, that should be subject to interpretation. In this case the correct interpretation is as follow. If the abstract series is continuous, the interpretation of its continuity as that of the empirical series means:

(1) the abstract continuous series is compared with the empirical series which is also continuous;

(2) each element of the abstract series is set in correspondence with a

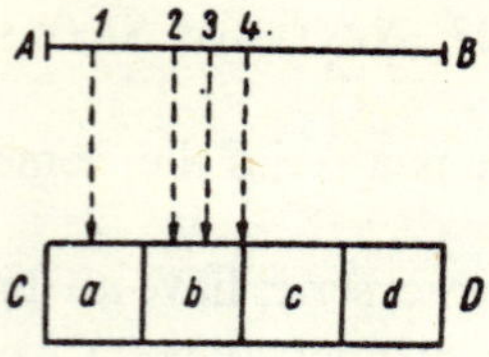

Fig. 4.

certain element of the empirical series, rather than with the break ('the empty space') between the elements of the empirical series, since the latter does not have such breaks.

Thus, if one 'projects' a point of the abstract series onto the empirical series, a certain empirical individual will turn up, which is an element of this series. Different points of the abstract series may correspond to the same individual of the empirical one.

In Figure 4 we present a segment AB the points of which form a continuous abstract series, and the objects *a*, *b*, *c*, *d* forming a continuous empirical series CD. As is indicated by the dashed arrows, if the series CD is an interpretation of the series AB, the projection of an arbitrary point of AB will meet one of the objects *a*, *b*, *c* and *d*. Points 2 and 3 are projected into the same object *b* while point 4 is projected into a joint of objects *b* and *c*, in other words, it is set in correspondence with two objects of the empirical series at once.

If an abstract series is interpreted in such a way that its every element is set in correspondence with an element of an empirical series, and its different elements correspond to different elements of the empirical series, then the continuous abstract series has no empirical interpretation.

Therefore, the interpretation of an abstract series is not just the interpretation of its elements. It also includes finding an appropriate interpretation for other terms having to do with the series, which does not contradict existing terminology for the empirical series.

Note incidentally, that the appearance of numerous paradoxes and contradictions in science is due to the confusion of abstract and empirical objects, and even if these are distinguished, one still comes across violations of the rules for interpretation of the former in terms of the latter.

28. Finite and Infinite Series

It is possible to distinguish between finite and infinite series from the point of view of the number of their elements and their length. The case of length is reduced to that of the number of the series elements, and this case will be treated in the next chapter.

Speaking of finite and infinite series, one must take into account the difference between empirical and abstract series. For instance, for an empirical series the following is valid: if the number of its elements is infinite, it has an infinite length. For abstract series this is not always the case. For example, the segment AB in Figure 4 contains an infinite number of points but is characterized by a finite length.

If an empirical series is limited, it has a finite number of elements and a finite length, and vice versa. An appropriate system of definitions may be constructed here, with quite interesting consequences.

Arbitrary assumptions, provided they are logically non-contradictory, are justified in the case of abstract series. Such, for instance, are the assumptions of infinitely long series. One may even imagine a series which is infinitely long, but has a starting and a final element (if, for instance, an infinitely large interval is assumed between its elements). On the other hand, in the case of empirical series such assumptions either remain empirically non-verifiable hypotheses or lead to abstract series that have no empirical interpretations. To give an example, an abstract series with infinitely long intervals between some of its elements has no empirical interpretation.

29. Structure

The following definition of structure appears to be intuitively acceptable: a structure is a cluster of individuals such that for any element a one can find another element b as well as a method of establishing order α such that $a > \alpha b$ or $b > \alpha a$. However, this kind of definition is appropriate only as an approximate one the purpose of which is to give the reader an outline of what will be discussed. It is impossible to carry out a rigorous argument with such a definition.

Let us first show how the primary terms of structures are introduced. They may be introduced in different ways, and, in particular, by the following:

An expression 'A structure which is formed by elements of cluster A with respect to a class of methods of establishing order B' will be regarded as a term whose meaning is identical to that of the term 'A cluster A of individuals such that for any of its elements a one can find its other element b and a method of establishing order α from a class B such that $a > \alpha b$ or $b > \alpha a$'.

In this definition the word 'structure' is defined as a part of complex term. In order to introduce it as a term, derived with respect to the primary terms of structures, one can make use of the general rules for introduction of terminology, in particular, of the rule of generalization.

The above definition is explicit. It makes it obvious that individuals of a certain cluster form a structure only with respect to some given methods of establishing order. If these methods are not specified, there can be no structure at all. And since the choice of these methods is up to the investigator, she has no right to inquire with respect to a certain cluster A whether it represents a structure or not, if the class of methods of establishing order B has not been specified (chosen) in advance. While a structure is a cluster, not every cluster is a structure (although this does not exclude the fact that for an arbitrary cluster of two or more individuals one can find certain methods of establishing order such that this cluster will be a structure).

The implicit definition of the terms of elementary structures is as follows: the elements of a cluster A form a structure with respect to a class of methods of establishing order B if and only if for any element a of this cluster one can find its other element b, and a method of establishing order α from a class B such that $a > \alpha b$ or $b > \alpha a$. By employing individual variables a and b and a variable for the methods of establishing order α, we can write the defining part (after the words 'if and only if') in the form:

$$(\forall a)(\exists b)(\exists \alpha)((a \in \mathrm{A}) \wedge (b \in \mathrm{A}) \wedge (c \in \mathrm{B}) \rightarrow ((a > \alpha b) \vee (b > \alpha a))).$$

In the implicit definition the word 'structure' is defined as a part of the statement. And additional logical operations are still needed for introducing the term 'structure' as an independent term. However, this is not necessary if this term is used in the approximate sense and one takes into account that in the argument one always implies terms denoting concrete structures (in the final analysis, the elementary terms of structures).

A special case of the cluster A described above is represented by a cluster consisting of individuals $a^1, \ldots, a^n (n \geqslant 2)$, and a special case of the class B, by the enumeration of the methods of establishing order $\alpha^1, \ldots, \alpha^m (m \geqslant 1)$.

As depicted in Figure 5, individuals a, b and c form a structure with respect to the methods of establishing order α^1 (the reference point for the order is x; the direction of the increase of order is indicated by the arrow) and α^2 (y and the arrow). But these individuals may form a structure with respect to other methods of establishing order, in particular, in a

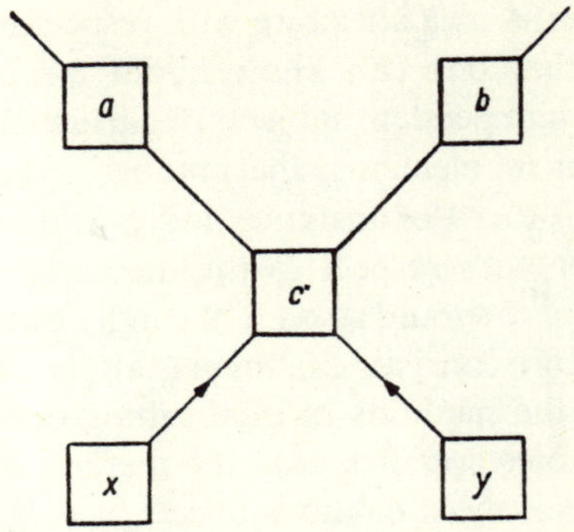

Fig. 5.

way shown in Figure 6. Here there is one reference point, but the order of *a*, *b* and *c* is established through projecting them on the straight line indicated by the large solid arrow. If one changes the method of projection (see the dashed arrows), a different structure of *a*, *b* and *c* will result, but it will still be a structure (it will be discussed below). On the other hand, the individuals *a* and *c* do not form a structure with respect to a method of establishing order, shown in Figure 6 by dashed arrows.

An ordered series, obviously, represents a structure. The simplest structure is that of two individuals and one method of establishing order.

The definition of a structure enables one to consider, or not to consider,

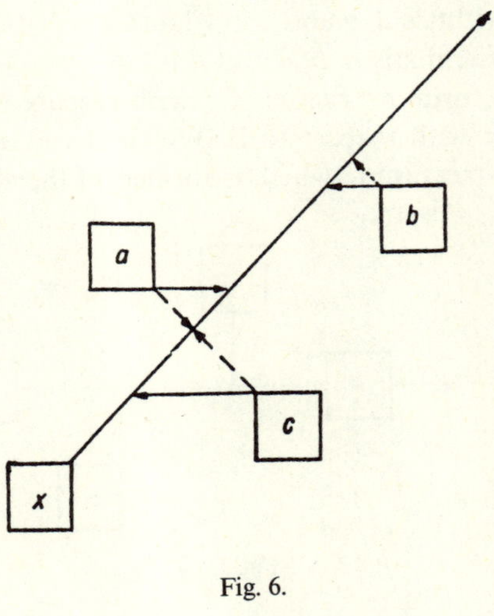

Fig. 6.

one or another cluster A as a structure with respect to B. But it does not contain everything that one can know about a structure. A structure may be studied as an independent subject. In particular, one can determine the intervals between its elements, the number of elements, their order with respect to each other (for instance, individuals *a* and *c* in Figure 6 are identical in order with respect to the method of establishing order shown by the dashed arrow), and so on. If the individuals of A are empirical individuals, the structure consisting of them is also an empirical individual, although we choose the methods of establishing order and may change them at will. This choice specifies only the method of description of the cluster A; the change of these methods affects just the way of description of the same cluster A, rather than the cluster itself. Everything that has been said about relations of order in general may be applied to structures, only in a more complicated form.

The use of structure-terms as subjects of statements ('predicating' of structures) does not yet mean that they can be associated with arbitrary predicates to form meaningful statements. There are predicates for which the conditions of their being ascribed to structures should yet be established (i.e., one should determine the meaning of their combinations with terms of structures). We are going to come across more examples of this sort.

Assume that a cluster A^1 forms a structure C with respect to B. Let A^2 be a cluster forming a structure D also with respect to B. If $A^1 \subset A^2$, we shall call C a substructure of the structure D with respect to B. If $\sim(A^2 \subset A^1)$, then C is the characteristic substructure of the structure D. The cluster of individuals *a*, *b* and *c* in Figure 7 represents a substructure of the cluster of individuals *a*, *b*, *c* and *d*.

Let a cluster A^1 form a structure C^1 with respect to B, and a cluster A^2 – a structure C^2 with respect to B. We shall call a join of structures C^1 and C^2 such a structure C which is formed of the elements of clusters

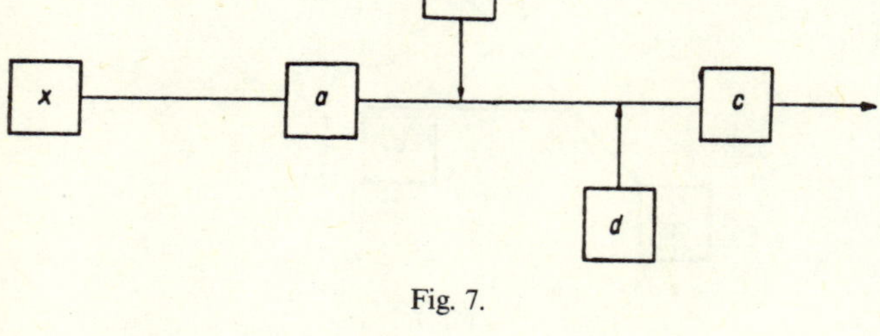

Fig. 7.

A^1 and A^2 with respect to B, the structures C^1 and C^2 being the substructures of C.

Let B be a class of methods of establishing order with respect to which the elements of the cluster A form a structure C. We shall say that a cluster of individuals D is inside C (is included in C) with respect to B if and only if all elements of D are between the elements of A with respect to B.

Depending on what the elements of structures represent – empirical or abstract objects – one can distinguish between empirical and abstract structures. Problems of empirical interpretation of the latter are still more complex than those of the empirical interpretation of abstract series.

An individual a is said to be inside a structure A with respect to β if and only of a is between the limiting points of A with respect to β.

An individual a is outside a structure A with respect to a class of methods of establishing order B if and only if for any pair b^i and b^k of its limiting points and for any α such that $\alpha \in B$, the following is valid:

$$(b^i > \alpha b^k) \rightarrow ((a > \alpha b^i) \vee (a < \alpha b^k)).$$

An individual a is not inside A with respect to B if and only if it either represents a limiting point of A or is outside of A with respect to B.

Corollaries: (1) if a is outside of A with respect to B, then there is a structure A* such that A is the characteristic substructure of A* with respect to B; (2) if a structure A is outside of B, B is outside of A; (3) if A is inside B, then B is not inside A.

If A is a structure of the type X with respect to a class of methods of establishing order α, then A is a structure of the same type X with respect to a class of methods of establishing order β such that $\alpha \subset \beta$. In particular, if A is a spatial (temporal) structure with respect to α, A is a spatial (temporal) structure with respect to β.

30. Existence of a Structure

The question of the existence of structures reduces (by virtue of the rules for construction of terminology) to that of the existence of structures denoted by the elementary structure-terms. And the existence of the latter depends on the existence of their elements and relations defined between them, satisfying the definition of the corresponding terms. For instance, a structure formed by the individuals a, b and c with respect to α, exists if and only if a, b and c exist, and there are order relations between them with respect to α that satisfy the definition of structure. The situation with the predicate of possibility is similar.

Thus, the assertion 'A structure does not exist (is impossible) if the individuals forming it do not exist (are impossible)' is a consequence of the definition of structure and the predicate of existence (possibility) as applied to relations in general and to structures in particular.

31. The Dimensions and Order of Structures

The question of the dimensions of structures reduces to that of the dimensions of individuals, intervals and series and of operations with corresponding quantities. Such are, as is well known, the operations for determining the areas and volumes of figures and bodies. We do not consider these topics. Let us just make a few remarks.

If one assumes that the dimensions of an empirical individual are always greater than zero, then the dimensions of an arbitrary empirical structure are also greater than zero. The dimensions of individuals are established in the same manner as the dimensions of a determinate structure formed by parts of these individuals.

The dimensions of a substructure A of a given structure B cannot exceed to dimensions of B. If an individual a is inside a structure A, its dimensions are not greater than the dimensions of A. If an individual a represents an element of a structure A, its dimensions are not greater than the dimensions of A.

Additional agreements should be made for establishing the order of structures. For instance, the following agreement is possible: a structure A exceeds in order a structure B with respect to α if and only if

$$(\forall a)(\exists b)((b \in \mathrm{B}) \wedge ((a \in \mathrm{A}) \rightarrow (a > \alpha b)),$$

where a and b are variables for terms of empirical individuals. A stronger definition will result if the determining part is given the form:

$$(\forall a)(\forall b)((b \in \mathrm{B}) \wedge (a \in \mathrm{A}) \rightarrow (a > \alpha b)).$$

As far as the first definition is concerned, a situation is possible when $\mathrm{A} > \alpha\mathrm{B}$, and A is inside B with respect to C such that $\alpha \in \mathrm{C}$. From the point of view of the second definition this is impossible.

32. Correspondence

To choose an object means to draw the attention to it in one way or another, to point to it, name it, represent it, etc. The choice of objects is an elementary cognitive operation, explained through examples.

The choice of two or more different objects represents their comparison. We must call attention to the fact that in the case of comparison of objects each of the objects compared may be chosen independently (of others). Since the chosen objects are different, an object is not compared to itself.

If a certain rule is accepted (established) for comparison of objects, we shall say that a correspondence between these objects is established. A correspondence is established only between different objects. More precisely, the notion of correspondence is defined for its concrete types. A correspondence of objects is considered established from the point of view of a given group of persons without whom it does not exist at all. Regardless of the persons establishing the correspondence of objects, the latter are not in correspondence by themselves.

The statement 'An object b corresponds to an object a' will be written in symbolic form as

$$a \Leftarrow b,$$

where $\Leftarrow$ is a two-place predicate 'the second corresponds to the first'.

If a and b represent individual terms, the predicate $\Leftarrow$ has the following meaning: $a \Leftarrow b$ if and only if $\sim(a \rightleftharpoons b)$, and it is decided that under certain specified conditions x one always chooses b after the choice of a. The expression 'Always after the choice of a, b is chosen' is equivalent to the expression 'If a is chosen, b is chosen after that'. Thus, the expression $a \Leftarrow b$ is but an abbreviated form of a more complex expression, containing a statement of the type $y \to z$.

If b is a general term, and a is an individual one, the predicate of correspondence is defined as follows:

$$(a \Leftarrow b) \equiv \text{Df.}\ (\forall\alpha)((\alpha \rightharpoondown b) \to (a \Leftarrow \alpha)),$$
$$(a \urcorner \Leftarrow b) \equiv \text{Df.}\ (\exists\alpha)((\alpha \rightharpoondown b) \wedge \sim(a \Leftarrow \alpha)),$$

where α is an individual variable (in other words, $a \Leftarrow b$ if and only if any individual which is b corresponds to a). If a is a general term, and b is an individual term, then the above definition will take the form:

$$(a \Leftarrow b) \equiv \text{Df.}\ (\forall\alpha)((\alpha \rightharpoondown a) \to (\alpha \Leftarrow b)),$$
$$(a \urcorner \Leftarrow b) \equiv \text{Df.}\ (\exists\alpha)((\alpha \rightharpoondown a) \wedge \sim(\alpha \Leftarrow b)),$$

(i.e., $a \Leftarrow b$ if and only if any individual, which is a, corresponds to b). If a and b are both general terms, then

$$(a \Leftarrow b) \equiv \text{Df.}\ (\forall\alpha)(\forall\beta)((\alpha \rightharpoondown a) \wedge (\beta \rightharpoondown b) \to (\alpha \Leftarrow \beta)),$$
$$(a \urcorner \Leftarrow b) \equiv \text{Df.}\ (\exists\alpha)(\exists\beta)((\alpha \rightharpoondown a) \wedge (\beta \rightharpoondown b) \wedge \sim(\alpha \Leftarrow b)),$$

where α and β are individual variables. For example, any specimen of the word 'table' corresponds to any specimen of tables.

It follows from the definitions that

$$(a \rightleftharpoons b) \to \sim (a \Leftarrow b) \wedge \sim (b \Leftarrow a).$$

33. Correspondence of Classes

The establishment of a correspondence of objects is practically appropriate in those cases when one deals with relations between classes of objects. As far as we see it, it is meaningful only for non-empty and non-intersecting classes. Our argument is based on the following considerations.

In the process of establishing correspondence between intersecting classes (having common elements), we discover that their common elements become doubled in one way or other. If an individual a is included both in a class A and in a class B, then in comparison of A and B as different classes (otherwise the comparison is impossible), a is regarded as a general term while the expressions $a\downarrow(a \in \mathrm{A})$ (i.e., an individual a is chosen, that is included in A) and $a\downarrow(a \in \mathrm{B})$ (i.e., an individual a is chosen, that is included in B) are considered as individual terms.

Consider the following example. Let a certain type of correspondence be established between the class of positive integers A and the class of even numbers B. A number 2 is included both in the class A and in the class B. But the point is that the expression 2 is not an individual term (2 is not the only individual denoted by the word 'two'. There are many individuals that represent 'two'). And they can be produced in any amount: 2, 2, 2, ... And they are all different (at least because they occupy a different place in space). The situation for all other numbers is similar. Thus, in establishing the correspondence between classes of positive integers and even numbers, one considers, actually, instead of A and B, quite different classes which are just their subclasses, namely: (1) a class, including only one specimen of the number one, only one specimen of the number two, etc.; (2) a class, including only one specimen of the number two, only one specimen of the number four, etc. The terms of these classes (A* and B*, respectively), unlike A and B, are general, and not individual. The numbers 'two', included in these classes, represent different individuals. Since arbitrary classes of this kind are considered, the explicit form of the resulting assertion becomes $(\forall\alpha)(\forall\beta)x$, where α and β are the variables for this kind of classes and x is the result of their comparison. Classes A* and B* are non-intersecting (disjunct) classes. On the other hand, classes

A and B are such that $B \subset A$. Thus, A contains a totally unlimited and unaccountable measure of units, two's, etc. The situation with B is similar.

Thus, we shall say that class B is set in correspondence [coordinated] with class A if and only if A and B do not intersect, are both non-empty, and each element of A is set in correspondence [coordinated] with at least one element of B. This is a definition of the predicate $\Leftarrow$ for class-terms. It follows from the definition:

(1) if at least one of A and B is empty, then $\sim(A \Leftarrow B) \wedge \sim(B \Leftarrow A)$;

(2) if A and B intersect, then $\sim(A \Leftarrow B) \wedge \sim(B \Leftarrow A)$;

(3) if $A \subset B$, then $\sim(A \Leftarrow B) \wedge \sim(B \Leftarrow A)$;

(4) $\sim(A \Leftarrow A)$.

Let a, b, c, d be individual variables; A and B, variables for classes; x in a statement 'A and B are non-empty and do not intersect', y a statement 'Each element of A is set in correspondence with one and only one element of B', i.e.,

$$(\forall a)(\exists b)(\forall c)((a \in A) \wedge (b \in B) \wedge (a \Leftarrow b) \wedge ((a \Leftarrow c) \wedge \\ \wedge (c \in B) \rightarrow (b \rightleftarrows c))),$$

z is a statement 'Different elements of A are coordinated with different elements of B', i.e.,

$$(\forall a)(\forall b)(\forall c)(\forall d)((a \in A) \wedge (b \in A) \wedge (c \in B) \wedge \\ \wedge (d \in B) \wedge (a \Leftarrow c) \wedge (b \Leftarrow d) \wedge \sim(a \rightleftarrows b) \rightarrow \sim(c \rightleftarrows d)).$$

We shall adopt the following definitions:

(1) a class B is in one-to-one correspondence with a class A (in the abbreviated form, $A \Lleftarrow B$), if and only if $x \wedge y \wedge z$, in other words,

$$(A \Lleftarrow B) \equiv \text{Df.}\ (x \wedge y \wedge z);$$

(2) classes A and B are in one-to-one correspondence (in the abbreviated form, $A \Leftrightarrow B$), if and only if $(A \Lleftarrow B) \wedge (B \Lleftarrow A)$, i.e.,

$$(A \Leftrightarrow B) = \text{Df.}(A \Lleftarrow B) \wedge (B \Lleftarrow A).$$

34. FUNCTIONS

If a general method is specified (established) by means of which one can determine for each element of Ka exactly which elements of Kb are in correspondence with it, we shall say that a method or a type of correspondence of the class Kb to the class Ka is specified (established), or that a function b of a is known.

Statements to the effect that b is a function of a will be symbolically written in the form

$$b \Leftarrow f(a),$$

where f denotes the way of correspondence (the type of function), and $\Leftarrow f$ is a two-place predicate 'the first one is a function f of the second'.

For the same Ka and Kb, one may establish different ways of correspondence that are fixed by the statements

$$b \Leftarrow f^1(a), \ldots, b \Leftarrow f^n(a),$$

where $n \geqslant 1$. For example, the correspondence of the class of odd numbers Kb to the class of even numbers Ka may be established through the equation $b = a + 1$ and through the equation $b = a \pm 1$. And if one adopts a decision to establish the above kinds of correspondence between the class Kb and the class Ka, all $b \Leftarrow f^i(a)$ are regarded as true.

35. Ordered States

From the point of view of science, it is important to fix the order relations between states. For instance, 'After the bell had rung, the dog started salivation', 'A magnetic field is produced around the conductor when the electric current is passing through it', etc.

A special case of statements, taking into account the order of events, are such statements as 'y in relation R to x (or to what is x)'. Their symbolic form will be

$$(\mathrm{R}x)y.$$

For example, 'y an hour after x'.

Further, there are statements of the type 'x and then y', 'x and before it y', 'x and to the left of it y', etc. These peculiar ordered conjunctions are frequently used implicitly and become confused with ordinary ones. We believe that it is exactly this confusion which is responsible for the exclusion of certain laws of classical logic in the 'logic of the micro-world'.

In the general case, ordered conjunctions have the form

$$x \wedge (\mathrm{R}x)y.$$

Similarly, ordered disjunctions are possible:

$$x \vee (\mathrm{R}x)y.$$

For these conjunctions and disjunctions the following assertions

$$x \wedge (\mathrm{R}x)y \rightarrow y \wedge (\mathrm{R}y)x,$$
$$x \vee (\mathrm{R}x)y \rightarrow y \wedge (\mathrm{R}y)x,$$

are not always valid. Therefore, one should not adopt the rules of commutation

$$x \wedge (\mathrm{R}x)y \vdash y \wedge (\mathrm{R}y)x,$$
$$x \vee (\mathrm{R}x)y \vdash y \vee (\mathrm{R}y)x,$$

for them. Only the following rules of commutation

$$(x \wedge (\mathrm{R}^1x)y) \wedge ((a\mathrm{R}^1b) \rightarrow (b\mathrm{R}^2a)) \vdash (y \wedge (\mathrm{R}^2y)x),$$
$$(x \vee (\mathrm{R}^1x)y) \wedge ((a\mathrm{R}^1b) \rightarrow (b\mathrm{R}^2a)) \vdash (y \vee (\mathrm{R}^2y)x,$$

are valid for them. For instance, from the statement '*x* and then *y*' it follows that '*y* and before it *x*', since from '*a* occurs after *b*' it follows that '*b* occurs before *a*'.

36. Conditional Statements with an Order-Relation

There are conditional statements of the form

$$x \rightarrow (\mathrm{R}x)y,$$

where R is the order-relation. For example, 'If an electric current is passed through a conductor, simultaneously (or immediately after that) a magnetic field is formed around the conductor'. Because of the presence of R, these statements acquire additional properties, in particular, the following ones:

$$(x \rightarrow (\mathrm{R}x)y) \dashv\vdash (\forall\downarrow x)((\mathrm{R}x)y,$$
$$(x \rightarrow (\mathrm{R}^1x)y) \wedge ((a\mathrm{R}^1b) \rightarrow (b\mathrm{R}^2a)) \vdash (\sim y \rightarrow (\mathrm{R}^2 \sim y) \sim x).$$

37. Functional Dependence

In principle, for any two non-empty classes K*a* and K*b* one may find a way of correspondence f such that $b \Leftarrow f(a)$. This is trivial, since we can agree, for instance, to set in correspondence to any *a*, a certain *b* or any *b*. This is trivial, since establishing the correspondence of classes represents a decision by the will of the persons dealing with classes.

There are cases, however, when the way of correspondence is forced by certain circumstances and is not chosen arbitrarily. This happens when there is a certain relation (dependence) between the individuals of one class and individuals of another class, regardless of whether they are studies or not. The statements, resulting in such cases, have the form

$$\downarrow(\mathrm{R}x)y \Leftarrow f(\downarrow x),$$

and they refer not simply to individuals but to states of individuals.

These statements are obtained from the following ones:

$$x^i \to (\mathrm{R}x^i)y^i$$

such that

$$(\downarrow x^i \rightharpoondown \downarrow x) \wedge (\downarrow y^i \rightharpoondown \downarrow y) \wedge (\downarrow y \Leftarrow f(\downarrow x)).$$

Here f is chosen in such a way that the condition

$$\vdash (\downarrow(\mathrm{R}x)y \Leftarrow f(\downarrow x)) \wedge x^i \to (\mathrm{R}x^i)y^i$$

is satisfied for any x^i and y^i such that $\downarrow x^i \rightharpoondown \downarrow x$ and $\downarrow y^i \rightharpoondown \downarrow y$. This kind of statement refers to the functional relation between states.

Consider an example. One observes states of objects a and b successively in time according to a certain attribute P. The magnitude of P is fixed for a, and then, after a certain time t, the magnitude of b with respect to this attribute is fixed. This is a type of relation R in a statement $(\mathrm{R}x)y$. A table of quantities a and b is compiled. The study of the latter allows us to establish that under certain approximations, assumptions, etc., the quantity b is always about two times greater than the quantity a. This represents a type of function f in the statement $\downarrow y \Leftarrow f(\downarrow x)$. Here the statements x^i and y^i refer to the quantity a, and, respectively, b, at a given time and within a range of quantities, established also through observation (this may be specified by a table).

A statement $\downarrow(\mathrm{R}x)y \Leftarrow f(\downarrow x)$ may be given a form of a conditional statement

$$x \to (\mathrm{R}x)y^*,$$

where f is introduced into y in such a way that a statement y^* results. Thus, in the above example the statement may be given the form: 'If a is characterized by a certain magnitude, then after a time t the object b will have a magnitude twice as large'.

38. Connections

Statements of the form

$$x \to (\mathrm{R}x)y,$$
$$\downarrow(\mathrm{R}x)y \Leftarrow f(\downarrow x),$$

such that the statement

$$x \rightarrow y,$$

is not logically provable (logically true) will be called statements referring to the connection between states $\downarrow x$ and $\downarrow y$, or to the physical entailment of $\downarrow y$ following $\downarrow x$. If a is such a statement, then $\downarrow a$ is a connection-term and what is denoted by this term represents a connection of states. The latter are elements of connection. We shall also say that the above states are connected.

Since a statement referring to the functional connection may be transformed into a conditional statement with a relation, in what follows we shall take a generalized form of statements referring to empirical connections: $x \rightarrow (\mathrm{R}x)y$, where y may contain a description of a certain function.

If in $x \rightarrow (\mathrm{R}x)y$ the relation R means the identity of coodinates for x and y, then Rx is omitted, and the statements acquire the form $x \rightarrow y$. But the relation Rx is implied in them in one way or other. For instance, in the statement 'If one passes the electric current through a conductor, a magnetic field arises around it' the same time, or the time immediately after the occurrence of the first event, is implied.

A statement A concerning connection is constructed in such a way that from statements x^i one could in a purely logical way obtain statements y^i such that $\downarrow x^i \in \mathrm{K} \downarrow x$ and $\downarrow y^i \in \mathrm{K} \downarrow y$, and x, y and Rx occur in A. In other words, A is constructed in such a way that

$$x^i \wedge (\downarrow x^i \in \mathrm{K} \downarrow x) \wedge \mathrm{A} \vdash (\mathrm{R}x^i)y^i$$

holds. Therefore, A may be regarded as a special rule for obtaining certain statements from others. But this is not a logical rule (i.e., it does not refer to the properties of the language); it deals instead with the properties of one or another object sphere.

Depending on the structure of x and y, we distinguish between simple and complex statements concerning connections. Corresponding to this, connections are also distinguished. Thus, in the statement $(a \wedge b) \rightarrow \rightarrow (\mathrm{R}(a \wedge b)c)$ the expression x represents a complex statement $a \wedge b$, consisting of two statements; in $a \rightarrow (\mathrm{R}a)(b \vee c)$ the expression y is complex.

The properties of complex statements referring to connections are defined through those of simple ones by applying special logical rules. These are, for instance, such rules as

$$(a \vee b \rightarrow (\mathrm{R}(a \vee b))c) \dashv\vdash (a \rightarrow (\mathrm{R}a)c) \wedge (b \rightarrow (\mathrm{R}b)c),$$
$$(a \rightarrow (\mathrm{R}a)(b \wedge c)) \dashv\vdash (a \rightarrow (\mathrm{R}a)b) \wedge (a \rightarrow (\mathrm{R}a)c),$$

etc. But, in addition, there are quite different rules possible here. We shall treat the issues of connections in greater detail in the fourth chapter.

39. Ordering of Classes

One must distinguish two types of ordering of classes. The first type of ordering (we shall call it definitive or conventional) consists in making a decision to ascribe the elements of a given class one or another kind of order (the ordinal number) and one or another order relation. For instance, a class A of objects is given, consisting of objects *a*, *b*, *c*, *d*. In the case of conventional ordering one makes a decision (in particular) to consider *a* as the first in order, *b* the second, *c* and *d* the third; in addition, one (implicitly or explicitly) assumes that *b* is superior in order to *a*; *c* and *d* are superior in order to *b*, *c* and *d* are identical in order. The method of ordering here is our decision itself (it will be denoted by α). This decision can be replaced by another, for example, by the following one (it will be denoted by β): we shall assume that $(b > \beta a)$, $(a > \beta c)$, $(a = \beta d)$.

The second type of ordering (we shall call it empirical) consists in agreeing to choose a reference point of order and a certain direction (for instance, 'trajectory') of the imagined motion of the observer, which is meant to realize the choice of the objects of a given class. For example, a class of houses in a certain area A is specified, and a decision is made to establish their order in the following way: (1) the starting point is a house or houses situated to the north of the others; (2) one moves to the right marking the houses by a certain method; (3) one moves from the marked house to the nearest non-marked external house. In this way, from the method of establishing order one cannot logically deduce the order and the relation of order between the elements of the class. This is found empirically.

To put it briefly, if a class of objects is specified and a conventional method of their ordering is adopted, one can determine in a purely deductive manner the order relations between the elements of the class. If, on the other hand, one specifies a class of objects and adopts an empirical method of their ordering, then the question of the order relations between the elements of the class is answered only by depending on additional information on their position (in particular, on observation data).

There may be cases of ordering of the elements of a class A such that a certain subclass B of this class is ordered conventionally, while another, say C, is ordered empirically. There also may be cases when a class A is divided into subclasses $B^1, \ldots, B^n$ ordered by different methods, both conventionally and empirically.

CHAPTER THREE

NUMBER, MAGNITUDE, QUANTITY

1. Numbers in the Language

Numbers are linguistic expressions. They play the role of subject-terms, parts of subjects, parts of predicates and quantifiers. It is just predicate-terms that numbers cannot be. Examples: (1) the number 'ten' in the statement 'The number of types of elementary particles is greater that ten' is a subject; (2) the numbers 'one' and 'two' in the statement 'The molecule of water consists of one oxygen atom and two hydrogen atoms' are parts of subjects; (3) the number '300,000' in the statement 'Light propagates with a velocity about 300,000 km/sec' is a part of the predicate; (4) the number 'three' in the statement 'For all three types of the airplanes, it is true that they reach supersonic velocity' is a quantifier.

Analysis of the logical properties of numbers leads to the analysis of positive integers through which all other kinds of numbers are defined. The definition of the logical properties of positive intergers is the subject of formal arithmetic which actually became a part of logic. And there is justification to it: the construction of formal arithmetic does not imply the existence of any developed mathematics, but it does imply explicitly formulated logic and operators exclusively by logical methods. Below we shall give a presentation of a formal arithmetic that is essentially different from the constructions of this kind known to us and agrees with our conception of logic.

In establishing the properties of numbers we shall make use of the predicates of superiority and identity in order. At the same time this will serve as the definition of the above predicates for number-terms and as the ordering of the class of numbers.

We shall omit description of the method of establishing order, under the following assumptions: (1) the order of numbers is established by convention with number theory, in particular, by that system of formal arithmetic which will be discussed below; (2) the order of numbers is established by our arbitrary willed decision. Thus, the expressions '*a* is greater than *b*' and '*a* is equal to *b*' for numbers, in our presentation, will be identical with the expressions '*a* is superior in order to *b*' and '*a* is identical in order to *b*', respectively.

We do not distinguish here between both forms of negation for numbers, in the sense that the following assertion

$$\sim (a > b) \vdash (a \urcorner > b)$$

holds. This means that it is up to us to establish complete order in a given class of numbers and to exclude cases of uncertainty. Because of this assertion, the following statements will be valid:

$$(a \urcorner > b) \dashv\vdash \sim (a > b),$$
$$(a \urcorner = b) \dashv\vdash \sim (a = b),$$
$$(a = b) \dashv\vdash \sim (a > b) \wedge \sim (b > a),$$

and in what follows we shall limit ourselves to the logical rules for the classical case.

2. Numbers as Terms

We consider numbers first of all as subject-terms with a quite definite value. From this point of view, the positive integers [natural numbers] represent elementary terms. They are not defined through other numbers. On the contrary, all other numbers are defined through them. They have the following meaning as subject-terms. Every individual may be named (denoted) by the term 'one' (1, 'unit'), i.e., every individual is a unit. Thus, the deeper mystery of the unit is that simply there is no mystery here. The complexity of the problem in this situation is associated with the complexity of realizing its triviality. Every pair of individuals may be named by a term 'two' (2, 'pair'). And, in general, every *n*-tuple of *n* individual is *n*.

The meaning of derived numbers as terms is established according to the general rules for introduction of derived terms. For instance, a^2 may be defined as an abbreviation for $a \times a$, and the latter as a sum of *a* terms each of them being equal to *a*. In this process one uses no special logical means of mathematics (logical means which are characteristic of mathematics alone do not exist at all!), but just the general means of logic. When these general logical means are given the form of the specific means of mathematics, this only leads to confusion and unnecessary complication introduced into quite simple things.

Each number exists not as one specimen but as many. The potential amount (quantity) of specimens of any number is unlimited. For instance, we can produce units (1, 1, 1,...), two's (2, 2, 2,...), etc. in a potentially unlimited amount. And in this respect the theory of numbers does not differ at all from other sciences constructing generalizations, and in particular including the sciences having to do with language. The theory of numbers produces an abstraction under the assumption that all specimens of a given number exhibit no difference whatsoever from the point of view of their properties which are under consideration, i.e., it neglects the place and the time of using numbers as well as their graphical and acoustic modifications.

This abstraction means that those who deal with numbers have the capability (1) to distinguish the forms of numbers as perceived objects (for instance, to distinguish between units and two's); (2) to identify different specimen of numbers of the same form (for instance, to establish that symbols 1, 1, 1, 1 'one' and 'unit' represent specimens of the same form of numbers, namely, specimens of the unit). This is a general scientific fact, and not a specifically mathematical one (it is similar to the way a biologist distinguishes between rabbits and wolves and identifies various specimens of rabbits). If it is established that the numbers a and b are specimens of numbers of the same form, one accepts $a \rightleftharpoons b$, in other words, one accepts that all cases of using numbers of the same form are equivalent.

One has to differentiate between the terms which denote numbers, and the numbers themselves in the role of terms. For instance, this is how the term 'unit' denoting any units as numbers is differentiated from the units as terms, denoting arbitrary individuals. In the first case the term 'unit' is not a number while in the second it is. The difference between these cases remains latent because in linguistic practice an arbitrary number of form a is also used as a term, denoting numbers a, but as terms, denoting numbers a, we use certain instances of numbers a. But these cases, incidentally, are not equivalent. For example, the expression 'a class of units' in which the word 'unit' is a term, denoting units, denotes a class that includes all units. And if in this expression the word 'unit' is a number, it represents a term of the class that includes all individuals.

If a and b are numbers, then

$$(a \rightleftharpoons b) \dashv\vdash (a = \alpha b),$$

where α is a certain system of the number theory ordering the numbers of the class that includes a and b.

Further, we can construct the systems of formal arithmetic following two paths, using in the alphabet either the numbers themselves or the terms, denoting numbers. We have chosen the first method, which, due to the abstraction mentioned above, possesses the same degree of generality as the second one, but is simpler and more understandable than it.

3. Base Arithmetic

By the term 'base arithmetic' (in abbreviated form, BA) we mean a system, which defines the properties of the positive [natural] integers. It results from adding the following to the system of logic:

The alphabet of BA:

(1) 1 – the unit;

(2) + – the number-forming operator of addition ('plus').

The definition of number in BA (in abbreviated form, NB);

(1) 1 is NB;

(2) if $a^1, \ldots, a^n$ ($n \geqslant 2$) are NB's, then $(a^1 + \cdots + a^n)$ is NB;

(3) something is NB only by virtue of (1)–(2).

The definition of a proper statement in BA (in abbreviated form, PS);

(1) if a and b are NB's, then $(a > b)$ is PS;

(2) if x is PS, then $\sim x$ is PS;

(3) if $x^1, \ldots, x^n$ ($n \geqslant 2$) are PS, then $x^1 \vee \ldots \vee x^n$ and $x^1 \wedge \ldots \wedge x^n$ are PS;

(4) something is PS only by virtue of (1)–(3).

The definition of a proper formula in BA (in abbreviated form, PFB):

(1) if x is PS, then $\vdash x$ is PFB;

(2) if x and y are PS, then $\vdash (x \rightarrow y)$ is PFB;

(3) something is PFB only by virtue of (1) and (2).

As an abbreviated form for

$$\sim (a > b) \wedge \sim (b > a)$$

we shall use the symbol

$$(a = b).$$

If $a, b, a^1, \ldots, a^n$ are NB, then all PFB of the form (A_1)–(A_3) are axioms of BA:

(A_1) $\vdash ((a + 1) > a)$,

(A_2) $\vdash ((a > b) \rightarrow ((a > (b + 1)) \vee (a = (b + 1)))$,

(A_3) $\vdash ((a^1 + \cdots + a^n) = c)$,

where c differs from $(a^1 + \cdots + a^n)$ only in the presence of brackets.

The definition of provable PFB (or of the theorem of BA; in abbreviated form, TBA):

(1) the axioms of BA are TBA;
(2) if x is PS, and x is provable in logic (represents a theorem of logic), then $\vdash x$ is TBA;
(3) if x and y are PS, and $\vdash (x \rightarrow y)$ is a theorem of logic, then $\vdash (x \rightarrow y)$ is TBA;
(4) if $\vdash (x \rightarrow y)$ and $\vdash x$ are TBA's, then $\vdash y$ is TBA;
(5) something is TBA only by virtue of (1)–(4).

We shall call the elimination procedure in BA the following process: from a in $(a > b)$ one eliminates the first unit (in the order of occurrence), then the first unit (in the order of occurrence) is eliminated from b; this operation is repeated for the subsequent units (in order of occurrence, if there are such) up to the moment when all units are eliminated from a and/or from b.

PS and PFB will be assigned the values 'true' and 'not-true' according to the following rules:

(1) PS $(a > b)$ is true if and only if the elimination procedure in a leaves at least one unit, while there remains nothing in b;
(2) PS $(a = b)$ is equivalent to $\sim (a > b) \wedge \sim (b > a)$;
(3) Ps x and PFB $\vdash x$ are equivalent;
(4) Ps $x \rightarrow y$ is equivalent to $\sim x \vee y$.

The rules for $\sim$, $\wedge$, $\vee$ are the usual ones.

The following assertions are valid:

(1) all TBA's are true (a consequence of this is the consistency of BA);
(2) if x is PS and true, then $\vdash x$ is TBA;
(3) if x and y are PS and y is true under the condition that x is true, then $\vdash (x \rightarrow y)$ is TBA.

The system BA is complete in the sense of assertions (2) and (3). If PFB $\vdash x$ is given, then, in order to establish whether it is TBA or not, it is sufficient to determine whether or not it is true. If PFB $\vdash (x \rightarrow y)$ is given, then, in order to establish whether it is TBA or not it is sufficient to determine whether y will be true or not for x being true. Thus, there is a decidability procedure for BA with respect to PFB.

We shall call the base arithmetic with zero (in abbreviated form, BAZ) a system which results from the following additions to BA. Addition to the alphabet: 0, zero. Addition to the definition of number: 0 is a number of BAZ. Additional axioms:

(A_4) $\vdash((a+0)=a)$,
(A_5) $\vdash((a+b)=(b+a))$.

Addition to the eliminating procedure: if *a* and/or *b* in PS do not contain units, the elimination procedure is fulfilled; if *a* and *b* do contain units, then... (continuation as in BA). Assertions (1)–(3) are valid in BAZ.

4. The Abbreviated Form of NB

The conventional system of recording numbers, from the logical standpoint, is a system of such definitions as

$$(2) = \text{Df.}\ (1+1),$$
$$(3) = \text{Df.}\ (2+1),$$

etc., and also of the following definitions:

$$ab = \text{Df.}\ ((a \times 10) + b),$$
$$abc = \text{Df.}\ ((a \times 100) + (b \times 10) + c),$$

etc., where $a, b, c, \ldots$ are digits $1, 2, 3, \ldots$.

5. The Universal Nature of Arithmetic

In the literature related to modern physics, one can quite frequently come across a statement to the effect that even the rules of arithmetic are not universal (i.e. are not always true). And to illustrate this with examples, authors cite cases when unification of objects takes place. From the logical point of view, these examples are similar to the case when a combination of two drops of mercury results in one drop. The rules $1+1=2$, $2+2=4$, etc. are assumed not to be valid in such cases. But this assertion arises because of a logical error. The rules of arithmetic $1+1=2$, $2+2=4$, etc. have nothing to do with the physical, chemical, etc., combination of objects. They just mean that the numbers $(1+1)$ and 2 are identical in order (and in meaning); the situation is similar with the numbers $(2+2)$ and 4; and so on. The rules of arithmetic are universal by the very method of their determination and have absolutely no exceptions.

6. Expanded Arithmetic (EA)

The alphabet of EA:

(1) 1 and 0 – the unit and the zero;

(2) $+$, $\times$, $-$, $/$, $\sqcup$, $\sqrt{\ }$ – the number-forming operators of addition, multiplication, subtraction, division, taking powers and roots.

Definition of number in EA (in abbreviated form, NE):

(1) 1 and 0 are NE's;

(2) if $a^1, \ldots, a^n$ $(n \geqslant 2)$ are NE's, then $(a^1 + \cdots + a^n)$ and $(a^1 \times, \ldots, \times a^n)$ are NE's;

(3) if a and b are NE's, then $(a - b)$, (a/b), a^b, ${}^a\sqrt{b}$ are NE's;

(4) something is NE only by virtue of (1)–(3).

Definitions of PS and PFB differ from corresponding definitions for BA only in that NE appears everywhere instead of NB.

If the letters a, b, c, $a^1, \ldots, a^n$ are NE's, then the PFB's given below represent the axioms of EA:

(A1) $\vdash ((a + 1) > a)$,

(A2) $\vdash ((a > b) \rightarrow (((a > (b + 1)) \vee (a = (b + 1)))$,

(A3) $\vdash ((a^1 + a^2 + \cdots + a^n) = c)$,

where c differs from $(a^1 + a^2 + \cdots + a^n)$ only by the presence of brackets.

(A4) $\vdash ((a + 0) = a)$,

(A5) $\vdash ((a + b) + (b + a))$,

(A6) $\vdash ((a \times 1) = a)$,

(A7) $\vdash (((a \times (b + c)) = ((a \times b) + (a \times c))$,

(A8) $\vdash ((a^1 \times a^2 \times \cdots \times a^n) = c)$,

where c differs from $(a^1 \times a^2 \times, \ldots, \times a^n)$ only by the presence of brackets.

(A9) $\vdash ((a \times b) = (b \times a))$,

(A10) $\vdash (a^1 = a)$,

(A11) $\vdash (a^{b+c} = (a^b \times a^c))$,

(A12) $\vdash (((a - b) > c) \rightarrow (a > (b + c)))$,

(A13) $\vdash ((a > (b - c)) \rightarrow ((a + c) > b))$,

(A14) $\vdash ((a > (b + c)) \rightarrow ((a - b) > c))$,

(A15) $\vdash ((a + c) > b) \rightarrow (a > (b - c)))$,

(A16) $\vdash ((a > b) \wedge (b > 0) \rightarrow ((a/b > c) \rightarrow (a > (b \times c)))$,

(A17) $\vdash ((a > b) \wedge (b > 0) \rightarrow ((c > a/b) \rightarrow ((c \times b) > a)))$,

(A18) $\vdash ((a > b) \wedge (b > 0) \rightarrow ((a > (b \times c)) \rightarrow (a/b > c)))$,

(A19) $\vdash ((a > b) \wedge (b > 0) \rightarrow (((c \times b) > a) \rightarrow (c > a/b)))$,
(A20) $\vdash ((a > b) \wedge (b > 0) \rightarrow ((\sqrt[a]{b} > c) \rightarrow (b > c^a)))$,
(A21) $\vdash ((a > b) \wedge (b > 0) \rightarrow ((c > \sqrt[a]{b}) \rightarrow (c^a > b)))$,
(A22) $\vdash ((a > b) \wedge (b > 0) \rightarrow ((b > c^a) \rightarrow \sqrt[a]{b} > c))$,
(A23) $\vdash ((a > b) \wedge (b > 0) \rightarrow ((c^a > b) \rightarrow (c > \sqrt[a]{b})))$.

The definition of a theorem in EA (in the abbreviated form, TEA) is similar to that in TBA.

Axiom (A4) gives an implicit definition of zero, Axioms (A6)–(A9) give implicit definitions of numbers with the multiplication operator, Axioms (A10)–(A11), that of numbers with the operator of taking powers, (A12)–(A15) with the subtraction operator, Axioms (A16)–(A19) with the division operator, Axioms (A20)–(A23) with the operator of taking roots. The limitations $(a > 0) \wedge (b > 0)$ in (A16)–(A23) are adopted in order to avoid the paradoxes of zero. They may be weakened, which is of no significance for us here.

We would like to point out to the difference between definitions such as $\vdash (\alpha = \beta)$, allowing us to replace numbers by other numbers, equal to them, and definitions of the form $\vdash (\alpha \rightarrow \beta)$ and $\vdash (\gamma \rightarrow (\alpha \rightarrow \beta))$, allowing us to replace statements with numbers by other statements. As a consequence, one can use definitions of the second kind to replace numbers, but only as a consequence.

The axioms of EA, given above, do not exhaust all the rules for operations with numbers of EA. They should be supplemented by the rules for all possible combinations of numbers. We shall not cite them here, since this is of no major significance. Let us give just a few examples:

$$\vdash ((a + b) - c) = (a + (b - c)),$$
$$\vdash ((a - b) - c) = (a - (b + c)),$$
$$\vdash (a/b + c) = (a + b \times c)/b,$$

etc. Taken together with the above rules for the simplest cases, these rules allow us to reduce arbitrary combinations of numbers in EA to numbers in BA.

7. Infinite Numbers

The system of formal arithmetic with infinite numbers results from making the following additions to EA:

Addition to the alphabet: $\sim$ – the operator of infinity.

The definition of an infinite number:

(1) if a is NB, then $\utilde{a}$ is an infinite number;
(2) if a is an infinite number, then $\utilde{a}$ is an infinite number;
(3) something is an infinite number only by virtue of (1) and (2).

In the definitions of PS and PFB we take into account the fact that infinite numbers are numbers in a given system.

Additional axioms:

(A*1) $\vdash (a > b)$,

where a is an infinite number, and n is NE.

(A*2) $\vdash ((a+b) \rightarrow (a+b))$,
(A*3) $((a > b) \rightarrow (a > b))$,
(A*4) $((a > b) \rightarrow (a > b))$.

Similar to the fact that the terms 'NB' and 'NE' are not numbers, the term 'infinite number' is not a number.

It is true that if an infinite number is added to a certain positive number, an infinite number will result, but it is not true that

$$\vdash ((a+b) = a),$$

where a is an infinite number. The addition of a certain positive number b to an infinite number a results in an infinite number, but a different one (if $b > 0$).

8. Formal Arithmetic with Metastatements

In the alphabet of BA and EA there are no variables for numbers. Thus, statements about numbers in BA and EA (i.e., metastatements) cannot be expressed in the language of these systems. For instance, a statement 'For every NB there is NB which is greater than it' cannot be written as a statement of BA. In order to be able to record such statements, one has to expand the language of BA and EA in an appropriate way.

Let us make the following additions to systems BA and EA, leading to systems MBA and MEA, respectively. The alphabet is supplemented by a list of variables for numbers. The following definition of the subject is adopted: (1) NB's (correspondingly, NE's) are subjects of MBA (correspondingly, of MEA); (2) the variables for NB (for NE) are subjects of MBA (MEA); (3) something is a subject of MBA (MEA) only by virtue of (1) and (2). In the definitions, everywhere the terms NB (NE) are replaced by the terms 'a subject of MBA (MEA)'. In the definitions of PS the following statement is added: if a is a subject, and x is PS, then $(\forall a)x$ and $(\exists a)x$ are PS.

Now we can write a metastatement, given above as an example, in terms of the language of MBA, as

$$(\forall p)(\exists q)(q > p),$$

where p and q are variables for NB.

The theorems of MBA (and MEA) are formulas, obtained according to the general rules for introduction of variables from the formulas of BA (and EA) and from those theorems (according to the general rules of logic) which are supplemented here by the rules for quantifiers. For instance, from the formula

$$\vdash ((a+1) > a),$$

we obtain a formula

$$\vdash (\exists p)(p > a),$$

by virtue of the rule (V^5), and then, by applying the rule (V^6), we arrive at the formula

$$\vdash (\forall q)(\exists p)(p > q).$$

9. Formal Arithmetic and the Theory of Numbers

The systems of formal arithmetic represent just one of the means of number theory, one which does not exhaust the totality of methods of the number theory in general. They are but the definitions of certain linguistic expressions of scientific language, and definitions by themselves do not yet form a science, regardless of the possibilities for deduction provided by them. We shall present our general considerations on this point, using as an example Fermat's Great Theorem (in abbreviated form, GFT).

GFT, formulated in the language of MEA, has the following form:

$$\vdash (\forall p)(\forall q)(\forall r)(\forall s) \sim (((s > 3) \vee (s = 3)) \wedge (p > 0) \wedge (q > 0) \wedge (r > 0) \wedge (p^s + q^s = r^s)),$$

where p, q, r, s, are the variables for NB. Both GFT and its negation are not theorems of the systems of logic, establishing the properties of the operators $\vee$, $\wedge$, $\sim$, $\exists$, $\forall$. This is trivial, since in the language of these systems the operators $+$, $\times$, $\cup$ are totally absent, and they cannot be defined as the derivates of the first (if expressions with the latter are regarded as abbreviations for expressions with the former). But both GFT and its negation are not provable in EA, for the reason that there are no variables for NE in the language of EA. There is only one way to establish whether GFT or its negation represents a theorem in MEA or not, namely, the following one: to find a quadruple of NB's, x, y, z, u such that in EA the formula

$$\vdash ((x > 0) \wedge (y > 0) \wedge (z > 0) \wedge ((u > 3) \vee (u = 3)) \wedge \wedge (x^u + y^u = z^u))$$

is provable, or to establish the impossibility of existence of such quadruples, in other words, to establish that for any NB's in EA a, b, c, n the formula

$$\vdash \sim ((a > 0) \wedge (b > 0) \wedge (c > 0) \wedge ((n > 3) \vee (n = 3)) \wedge \wedge (a^n + b^n = c^n))$$

is provable. In the first case one has to find such a quadruple of NB's in some way, and to be able to write it, while in the second case it is necessary to determine the method of individual analysis of all such quadruples of NB's and to prove a certain metatheorem. But this represents a research problem in the frame of an entire science, for the solution of which one would be justified to mobilize any admissible means. It is a heuristic problem, and not that of formal inference within the frame of EA. Certainly, whatever four concrete NB's x, y, z, u we consider, by virtue of the completeness and decidability of BA and by virtue of the reducibility of expressions containing the operators of multiplication, and taking powers to the numbers of BA, we can establish whether the formula $\vdash (x^u + y^u = z^u)$ is a theorem of EA or not. But this represents only a part in the solution of the problem under consideration, and not the whole solution. And only after this problem is solved shall we be able to say whether GFT, written in the language of MEA, is a theorem of MEA or not.

Thus, the question of whether or not for every formula of MEA one can determine its provability in this system, is just a reformulation of another problem: whether or not every statement about NE may be proved or disproved in arithmetic as a science. But in the case of science as the whole, the very formulation of this problem is meaningless for the reason that it cannot be stated accurately: for arithmetic as a science there is no precise definition of its statement that would be similar to the definition of PS and PFB in BA and EA.

10. Number-Terms

In systems of number theory such terms as 'a number in BA', 'a number in EA', etc., or in general, the terms 'A number in S', where S is the name of a certain system in number theory, are defined. Their generalization is a 'number' term, introduced according to the following scheme: if *a* is a number in a certain system of the number theory, then *a* is a number. This kind of terms is not a number. But it denotes numbers, i.e., they are number terms.

By making use of the rules of limitation, one can construct, of such terms, terms of the form $a{\downarrow}\mathrm{P}$ and $a{\downarrow}x$, where *a* is a 'number' term or 'number in S', P is a certain limiting attribute of numbers (for example, 'can be divided by two without a remainder'), *x* is a statement containing number terms. In correspondence with the general rules for definition, one introduces the derived terms, the defining part of the definition of which contains number terms (for instance, 'a prime' = Df. 'A number in BA that can be divided without a remainder only by itself and one'). This is how the diversified terminology of numbers is formed.

From number terms we can construct statements with predicate orders, which do not always represent proper statements of the systems in the number theory. For example, a statement 'Some numbers are greater than 100' is constructed of the terms 'number' and '100', but it is not PS in BA, EA, etc. These statements are governed by the general rules of the logic of order (i.e., of the section of logic concerned with the properties of order predicates), the logic of quantifiers, and so on. The role of the methods of establishing order for such statements belongs to systems in number theory and, in general, to the agreements adopted for relations between numbers.

11. The Existence of Numbers

There is no single definition for the predicate of existence when we deal with arbitrary number-terms. It is defined in relation to the method of their construction.

First of all, we must distinguish here between the existence of numbers as special objects, and the existence of objects, denoted by number terms. Thus, if in a certain way we introduce into the alphabet of the number theory a number 0, this means that we accept the existence of 0: it can be seen and can be written. But the object, denoted by the term 0, does not exist.

The problem of the existence of numbers is solved, depending on one or another system by which to define a certain class of numbers. For the system EA the following assertions hold, forming an implicit definition of the predicate E for NE,

(1) $\vdash E(m1)$,
(2) $\vdash E(m0)$,
(3) $\vdash (a^1 \rightharpoonup NE) \wedge \ldots \wedge (a^n \rightharpoonup NE) \rightarrow E(m(a^1 + \cdots + a^n))$,
(4) $\vdash (a^1 \rightharpoonup NE) \wedge \ldots \wedge (a^n \rightharpoonup NE) \rightarrow E(m(a^1 \times \cdots \times a^n))$,
(5) $\vdash (a \rightharpoonup NE) \wedge (b \rightharpoonup NE) \rightharpoonup E(m(a - b))$,
(6) $\vdash (a \rightharpoonup NE) \wedge (b \rightharpoonup NE) \rightarrow E(m(a/b))$,
(7) $\vdash (a \rightharpoonup NE) \wedge (b \rightharpoonup NE) \rightharpoonup E(m(a^b))$,
(8) $\vdash (a \rightharpoonup NE) \wedge (b \rightharpoonup NE) \rightarrow E(m(\sqrt[a]{b}))$,
(9) If $\vdash x$ is TBA, then $\vdash E(m(a^1, \ldots, a^n){\downarrow} y$, where $n \geqslant 1$; $a^1, \ldots, a^n$ are terms 'NE'; y is formed of x through substitution of $a^1, a^2, \ldots, a^n$ in place of one or more occurrences of a certain NE.
(10) In the remaining cases NE does not exist,

Since for any NE's a and b in EA the formula

$$\vdash (b/0 = a)$$

is unprovable, then, according to (10)

$$\vdash \neg E(NE{\downarrow}(b/0 = a)).$$

This does not mean, however, that $m(b/0)$ does not exist. By virtue of (6) and the definition of NE, we have

$$\vdash E(m(b/0)).$$

Since in EA the formula

$$\vdash (3^2 + 4^2 = 5^2)$$

is provable, according to rule (9),

$$\vdash E(m(a, b, c){\downarrow}(a^2 + b^2 = c^2)),$$

where a, b, c are terms of NE (this statement reads as follows: 'There is a triplet of NE's such that the sum of the squares of the first and the second one equals to the square of the third').

Since (-1) is not NE, $\vdash \neg E(m(-1)$, $\vdash \neg E(m\sqrt{-1})$, etc. in EA. But if in a certain system of the number theory, according to the definition of the numbers of this system, $(-a)$ is a number, then with respect to this system one should adopt the assertion $E(m(-a))$. And then $\vdash(m(-1))$, $\vdash E(m\sqrt{-1})$,etc. will be true.

But if in the case of a certain number a it is true that $E(ma)$, then it does not follow from this that there is an object, denoted by the number-term a, i.e., it does not follow that $\vdash E(a)$. In other words, the assertion

$$\vdash E(ma) \rightarrow E(a)$$

does not represent the logical truth. For instance, from the fact that $\vdash E(m0)$ it does not follow that $E(0)$; from $\vdash E(m(-1))$ it does not follow that $E(-1)$, and so on. Frequent cases of misunderstanding arise because of the confusion of these different aspects of existence.

The situation with the predicate of possibility is similar. Definitions can be constructed, which combine the possibility and existence predicates.

12. Number as a Part of a Subject

Statements containing numbers as parts of subjects represent just linguistic transformations of statements in which numbers are independent subjects. For example, the statement 'The molecule of water is formed of two hydrogen atoms and an oxygen atom' is a substitute for the statement 'The molecule of water is formed of a hydrogen atom, another hydrogen atom, and an oxygen atom'. In general, the expression 'n objects a' in the sense 'a combination of n objects a' is but an abbreviation for the expression 'a combination of the object a, another object $a, \ldots, n$th object a'.

13. Quantity

The symbol of quantity represents a linguistic expression, containing a certain number α and a linguistic expression β which fixes the method of obtaining α (i.e., the units and the method of measurement). The predicates of quantities are predicates with the structure

$$P_{\alpha\beta}$$

where P is the name of the attribute ('weight', 'velocity', 'length'), α is a number ('10', '100'), and β is the linguistic expression, mentioned above ('kg', 'km/sec', 'cm').

Predicates of the form $P_{\alpha\beta}$ result from abbreviation of complex expressions. The establishment of the rules for the corresponding operations constitutes the concern of the sciences in which these predicates are used. But there are certain general properties to these predicates that can be expressed in the language of logic.

Let a be a number occurring in a statement x, β a system of the number theory in which a is a number and in which the properties of such numbers are defined, p a variable for the numbers of this system, $x(a/p)$ an expression, formed of x through replacement of a with p everywhere where a occurs in x. We shall adopt the following rule for removing numbers from the statements:

(A1) $\vdash ((p\downarrow x(a/p)) = \beta a)$,

(this formula reads as follows: the number p is such that $x(a/p)$ equals a with respect to β).

From (A1) the following consequences for quantities result:

(T1) If $P_{\alpha\gamma}(a)$ and $P_{\beta\gamma}(a)$, then the relation between a and b with respect to the attribute P is identical to the relation between numbers α and β.

For instance, if $\alpha = \beta$, then $a = Pb$; if $\alpha > \beta$, then $a > Pb$.

Such expressions as 'The quantity of a with respect to P is α' (for example, 'The weight of a equals 10 kg') are formed as a result of linguistic transformations of expressions of the sort $P_{\alpha\beta}(a)$ or $p\downarrow x(a/p) = a$. In addition to (A1), the following rules hold for quantities:

(A2) $\vdash P(a) \rightarrow (P_{\alpha\beta}(a) \rightarrow (\alpha > 0))$,
(A3) $\vdash P_{\alpha\beta}(a) \wedge (\alpha \rightarrow 0) \rightarrow P(a)$,
(A4) $\vdash \neg P(a) \rightarrow (P_{\alpha\beta}(a) \rightarrow (\alpha = 0))$,
(A5) $\vdash P_{\alpha\beta}(a) \wedge (\alpha = 0) \rightarrow \neg P(a)$.

For instance, if a body moves, the velocity of its motion is greater than zero, and vice versa; if a body is not moving, the velocity of its motion is equal to zero. The assertions (A2)–(A5) represent a part of the implicit definition of quantity.

The heuristic principle, according to which any empirical attribute can be measured, is completely legitimate, since, when necessary, for any 'qualitative' ('non-measurable') empirical attribute we in practice find a method of measurement. Consider, for instance, an attribute denoted by the word 'consciousness'. It can be measured by a number x/y, where y is the total number of the deeds of a man to whom the predicate 'consciousness' is applicable, and x is the number of this man's deeds when it is believed that he has manifested consciousness (his remaining deeds are considered immoral).

The predicates of quantities are divided into elementary and derived, in the following sense. If there are no such statements $x^1, \ldots, x^n$ in which $P_{\alpha\beta}$ does not occur and for which it is true that

$$\vdash (x^1 \wedge \ldots \wedge x^n) \rightarrow P_{\alpha\beta}(a),$$

then the predicate $P_{\alpha\beta}$ is elementary. If, on the contrary, there are such $x^1, \ldots, x^n$, then $P_{\alpha\beta}$ is derived.

Statements containing the elementary quantity predicates result from certain special empirical operations, called measurement. Statements with derived quantity predicates are obtained from other statements containing quantity predicates according to some special rules of an extralogical nature. Thus, as a result of the measurement of a distance α covered by a body A, and a time β spend on it we obtain statements with the elementary quantity predicates 'A has covered a distance of α km', and 'A has spent β seconds on it'. And by applying the rule for computation of velocity $v = s/t$ we obtain from these statements an expression with the derived quantity predicate 'A moved with an average velocity of α/β km/sec'.

Such distinguishing of quantity predicates is an extralogical distinguishing, which sometimes depends on specific features of the situation, on variations of the methods of measurement and computation, etc.

Finally, let us adopt the following definitions:

(1) A quantity a is minimal with respect to a class of quantities A (in abbreviated form, $1a = \min A$) if and only if $(\forall b)((b \in A) \wedge (a \in A) \rightarrow (1a \leqslant 1b))$.

(2) A quantity a is maximal with respect to a class of quantities A (in the abbreviated form, $1a = \max A$) if and only if $(\forall b)((b \in A) \wedge (a \in A) \rightarrow (1a \geqslant 1b))$.

(3) $(1a > \min A) \equiv \text{Df.}\ (\exists b)((a \in A) \wedge (b \in A) \wedge (1b < 1a))$.

(4) $(1a < \max A) \equiv \text{Df.}\ (\exists b)((a \in A) \wedge (b \in A) \wedge (1b > 1a))$.

14. The Degrees and Range of Truth

Statements $P_{\alpha^1\beta}(a), P_{\alpha^2\beta}(a), \ldots$ may differ only in numbers $\alpha^1, \alpha^2, \ldots$, but may all be regarded as true with respect to the same individuals a. For example, the expressions 'a weighs 10 kg', 'a weighs 9.99 kg', 'a weighs 10.1 kg', etc., may be considered as equally acceptable from a certain point of view. A class of individuals a, with respect to which $P_{\alpha^i\beta}(a)$ is true, is the domain of truth for $P_{\alpha^i\beta}(a)$. And the ranges of the values of numbers α^i in true $P_{\alpha^i\beta}(a)$ form the truth range of such statements.

Further, it is known that the quantities of attributes of objects are measured with a greater or lesser degree of accuracy or approximation. The statements $P_{\alpha\beta}(a)$ do not contain knowledge of this degree. They represent specific truth values, degrees of truth or degrees of approximation of the truth. There are no logical methods whatsoever for measurement of degrees of truth. But if they are measured in a certain way, some logical rules are valid for them (for the sake of brevity, we shall denote by the symbol $v(a)$ the expression 'the degree of truth of a'), for instance, such as

$$\vdash v(x \wedge y) \leqslant \min(v(x), v(y)),$$
$$\vdash v(x \vee y) \geqslant \max(v(x), v(y)),$$
$$(x \vdash y) \wedge (v(x) = \alpha) \vdash (v(y) \leqslant \alpha).$$

Increase of the degree of truth cannot expand a statement's domain of truth, but it can narrow it. Decrease of the degree of truth does not narrow the domain of truth. Variations of the degrees of truth take place inside the truth range.

15. Measurement and Definition

In practice we come across definitions of terms constructed according to the following scheme: if the application of measurement method A to objects a results in B, we shall say that $P_{\alpha\beta}(a)$. For instance, if a body a is put on a scale x (the method of measurement), and the scale indicator takes a position 1 (the result of measurement), then we shall say that a weighs 1 kg.

In definitions of this kind we come across predicates $P_{\alpha\beta}$ on the whole, rather than P and β separately. The meaning of P may be established through generalization, by making use of the rule:

$$\vdash (P_{\alpha\beta} \rightharpoonup P).$$

16. Number-Quantifiers

Expressions containing numbers as quantifiers are replaced by expressions with numbers as subjects, according to rule (A1) of Section 13. For instance, $(Na)x$, where $N \geqslant 0$, is replaced by 'A number of objects $a{\downarrow}x = N$'.

17. Amount

Expressions 'The amount [quantity] of elements of a class (cluster)' and 'The cardinality of a class (cluster)' will be used by us as synonyms. Since they are defined in the same way for classes and clusters, our discussion will be limited to the case of classes.

To define the expression 'The cardinality (the amount of elements) of a class' means, first of all, to define such expressions as 'A class *a* is identical in cardinality to a class *b*', 'a class *a* is superior in cardinality to a class *b*', 'a class *a* is inferior in cardinality to a class *b*', and their negations. To this end we shall use the predicates $=m$, $>m$, $<m$. On the other hand, in order to define these predicates, one has previously to define the standard classes of numbers. For this purpose, we shall also need the correspondence predicates, which have already been defined.

18. The Standard Classes of Numbers

D1. The definition of a standard NB:

(1) 1 is a standard NB;
(2) if *a* is a standard NB, then $(a+1)$ is a standard NB;
(3) something is a standard NB only by virtue of 1 and 2.

D2. A class of NB's will be called a standard class of NB's (in the abbreviated form, SCN), if and only if it includes only the standard NB's, and for every standard NB a the following is valid: SCN includes one and only one *a*.

To put it differently, SCN includes one and only one 1, one and only one NB $(1+1)$, one and only one NB $((1+1)+1)$, etc. for all NB's.

T1. As follows from D1 and D2, every SCN represents an ordered series, having the starting element 1 and having no final element.

D3. A class, including one and only one standard NB *n* and one of each standard NB's *a* such that $\vdash(n>a)$ and no other elements, will be called a standard *n*-class (in the abbreviated form, *n*-SCN).

T2. As follows from D3, *n*-SCN is 1 or (for $n>1$) represents an ordered series with the starting element 1 and the final element n (in other words, n-SCN is finite with respect to the number of terms).

In what follows, in order to illustrate one general statement, we shall need definitions of standard classes of even and odd numbers in base arithmetic. Let us cite these definitions.

D4. A class of NB's, including all even standard NB in the amount of one of each kind, and only such NB's, will be called a standard class of even NB's.

D5. A class of NB's, including all odd standard NB's in the amount of one of each kind and only such NB's, will be called the standard class of odd NB's.

19. The Cardinality of Classes of Numbers

D1. The cardinality of SCN is such as indicated in the defining part of D2 of the preceding section.

D2. The cardinality of n-SCN is such as indicated in the defining part of D3 of the preceding section.

The cardinality of other classes may be defined through that of SCN and n-SCN (if one says, for instance, that the cardinality of a class A equals that of SCN). But the cardinality of SCN and n-SCN is not defined through the cardinality of other classes. And in this sense the expressions 'The cardinality of SCN' and 'The cardinality of n-SCN' are primary. Definitions D1 and D2 represent just the identification of a certain (namely, quantitative) point of view towards classes, defined in D2 and D3 of the preceding section.

A1. The relation of cardinalities of a-SCN and b-SCN is identical to relation of a and b.

D3. The cardinality of a class of NB's is such that it includes all NB's.

A class of NB is not an ordered series, since it includes at least two NB's with the same order, for instance, $((1 + 1) + 1)$ and $(1 + (1 + 1))$. If one considers the fact that the elements of a class form an ordered series as a criterion for its countability, then the class of NB's is not countable. But each of SCN and n-SCN is a countable class.

If the cardinality of classes is expressed by numbers, this is only a linguistic transformation of statements, comparing the cardinality of these classes with that of classes of numbers. In the implicit form this can be written as 'The cardinality of A equals α' $\equiv$ Df. 'A is equal in cardinality to a class β', where the class β may be a class of NB's, SCN, n-SCN, etc. But from this kind of definitions it follows that if the cardinality of A equals α, and the cardinality of B equals β, then the relation of cardinalities of A and B is identical to the relation of numbers α and β (in the sense, similar to A1).

20. Comparison of Cardinalities of Classes

As an abbreviation for $(x \subset y) \wedge (y \subset x)$, we shall use the symbol $(x \subset \supset y)$.

Let us adopt the following assertions, defining the predicates $>m$ and $=m$ for classes:

(A1) $\vdash (\mathrm{A} \subset \mathrm{B}) \rightarrow (\mathrm{A} \urcorner > m\mathrm{B})$,

(A2) $\vdash (\exists a)(a \in \mathrm{A}) \wedge (\forall a) \sim (a \in \mathrm{B}) \rightarrow (\mathrm{A} > m\mathrm{B})$,

(A3) $\vdash (\mathrm{A} \Lleftarrow \mathrm{B}) \rightarrow (\mathrm{A} \urcorner > m\mathrm{B})$,

(A4) $\vdash (\mathrm{A} \subset \mathrm{C} \cup \mathrm{D}) \wedge (\mathrm{B} \subset \supset \mathrm{C} \cup \mathrm{E}) \rightarrow ((\mathrm{D} > m\mathrm{E}) \leftrightarrow (\mathrm{A} > m\mathrm{B})) \wedge ((\mathrm{D} = m\mathrm{E}) \leftrightarrow (\mathrm{A} = m\mathrm{B}))$,

(A5) $\vdash x \rightarrow ((\mathrm{A} > m\mathrm{B}) \downarrow (\exists c)(\exists d)((c \subset \mathrm{A}) \wedge (c \langle \equiv \rangle \mathrm{B}) \wedge (d \in \mathrm{A}) \wedge \sim (d \in c)))$,

where x is 'A and B are not empty and do not intersect'; c is a variable for classes; d is an individual variable.

From (A1)–(A5) follows:

(T1) $\vdash (\mathrm{A} \subset \supset \mathrm{B}) \rightarrow (\mathrm{A} = m\mathrm{B})$,

(T2) $\vdash (\mathrm{A} \Leftrightarrow \mathrm{B}) \rightarrow (\mathrm{A} = m\mathrm{B})$,

(T3) $\vdash (\forall a) \sim (a \in \mathrm{A}) \wedge (\forall a) \sim (a \in \mathrm{B}) \rightarrow (\mathrm{A} = m\mathrm{B})$,

(T4) $\vdash (\forall a)(a \in \mathrm{A}) \wedge (\forall a)(a \in \mathrm{B}) \rightarrow (\mathrm{A} = m\mathrm{B})$.

Further, from (A1)–(A5) we obtain the following important corollaries that do not always agree with the conventional way of reasoning.

T5. A class of subclasses of a class of classes cannot be superior in cardinality to the class of classes (since the former is a subclass of the latter).

The well-known assertion of set theory, which is opposite to T5, cannot be deduced in our formalism because one cannot apply correspondence predicates to the classes, mentioned in T5.

T6. The class of even numbers is not superior in cardinality to the class of NB's. A similar assertion is valid for the class of odd numbers.

T7. A standard class of even (odd) NB's that does not intersect with SCN is identical to the latter in cardinality (since there is one-to-one correspondence between them).

T8. Any two non-intersecting SCN's have equal cardinality.

T9. Any two non-intersecting *n*-SCN's have equal cardinality.

T10. Any two non-intersecting standard classes of even (odd) NB's have equal cardinality.

21. Other Definitions

D1. The definition of the predicate 'actually finite' for classes: (1) if

$$(\exists a)((a \in \mathrm{A}) \wedge (\forall b)((b \in \mathrm{A}) \rightarrow (a \rightleftarrows b)),$$

where *a* and *b* are individual variables, than A is actually finite;

(2) if A and B are actually finite, then $\mathrm{A} \cup \mathrm{B}$ is actually finite;
(3) if $\mathrm{A} \subset \supset \mathrm{B}$, and B is actually finite, then A is actually finite;
(4) if $\mathrm{A} \Leftrightarrow \mathrm{B}$, and B is actually finite, then A is actually finite;
(5) a class is actually finite only by virtue of (1)–(4).

D2. The definition of the predicate 'actually infinite' for classes: (1) if

$$(\forall \beta)((\beta \subset \mathrm{A}) \wedge \sim(\mathrm{A} \subset \beta) \wedge \mathrm{X} \rightarrow (\exists c)((c \in \mathrm{A}) \wedge \sim (c \in \beta)),$$

where β is a variable for classes, *c* is an individual variable, X is 'is actually finite', then A is actually infinite;

(2) if A is actually infinite, then $\mathrm{A} \cup \mathrm{B}$ is actually infinite;
(3) if $\mathrm{A} \subset \supset \mathrm{B}$, and B is actually infinite, then A is actually infinite;

(4) if A$\Leftrightarrow$B and B is actually infinite, then A is actually infinite;

(5) a class is actually infinite only virtue of (1)–(4).

The definitions of the predicates of existential and potential finiteness and infiniteness of classes are obtained from D1 and D2 by means of (1) substitution of the word 'existential' or, correspondingly, 'potential' in place of the word 'actual'; (2) replacement of the expression $(\exists a)x$ by the expression $E(a\downarrow x)$ or, correspondingly, by $M(a\downarrow x)$; (3) replacement of the expression $(\exists c)x$ by the expression $E(c\downarrow x)$ or, correspondingly, by $M(c\downarrow x)$.

The predicates 'not-finite' and 'not-infinite' (in the sense of actual, existential and potential) are defined according to the general rule:

$$(\tilde{P})(a) \dashv\vdash \sim P(a).$$

Since from $\sim P(a)$ it does not follow that $\urcorner P(a)$, the predicates 'not-finite' and 'infinite' are not identical. Similarly, the predicates 'finite' and 'not-infinite' are not identical.

22. Reducibility to Logic

Because of the definability of numbers, in the final analysis, through 1 and +, because of the fact that 1 as a term is equivalent to the term 'object', and because of the reducibility of all functions of numbers in the language to their functions as subject-terms, the following assertion holds: many problems in the methodology of science, which deal with the usage of numbers, quantity terms, measurement, etc., can be reduced to problems which are formulated in the language of logic and solved by logical methods. This is true, certainly, under the condition that we ignore all those problems which arise as a consequence of combining linguistic elements that are simple from the logical point of view, and the introduction of corresponding abbreviations.

Logical analysis of the language of science can, without any ill effects on the heart of the matter, eliminate the fact of number usage and consider the language only on such a plane that the results of mathematics as a science as well as those of the mathematization of the concrete scientific disciplines play no role whatsoever; this is possible since the use of numbers and operations with them does not by itself generate any new rules of logic nor does it remove the 'old' ones. The 'logical apparatus' of science, in general, does not depend on the fact of the use of numbers and of operations with them.

Finally, the rules ('laws') of number theory treat numbers as particular linguistic expressions. Hence, they provide a possibility of predictions only within their proper domain – in the field of number theory. They provide no means of prediction in the concrete sciences, by using numbers. The method of prediction in such sciences is just the set of assertions of these sciences themselves. An illusion to the effect that the rules of number theory are means of prediction arises because the assertions of the concrete sciences are implicitly associated with the rules of number theory. Consider, for instance, the following statements: (1) 'if a is increased *n* times, then *c* is increased *m* times'; (2) 'If *b* is increased *k* times, then *c* is decreased 1 times'. It appears that from (1) and (2) by virtue of the rules of mathematics, that one can derive a statement: (3) 'If *a* is increased *n* times, and *b* is increased *k* times, then *c* will change by a factor of $m/1$'. But in reality, in order to obtain (3), one has to make a non-mathematical assumption (4) such that from (1), (2), (4) one can derive (3). This may be, for instance, an assumption according to which the nature of the effect of *a* on *c* (and of *b* on *c*) is independent of whether *b* acts simultaneously on *c* (or respectively, *a* acts on *c*) or not. And in all those cases when it appears that a certain concrete scientific fact was predicted in a purely mathematical manner one can also discover some implicit assumptions concerning the objects studied by this science. This statement of ours is based on logical analysis of numbers as linguistic phenomena, characterized by *a priori* validity and does not need to be illustrated by examples of modern errors related to it.

23. A Remark on the Class of Positive Integers

Let x and y be the variables for positive integers. In metaarithmetic the formula

(1) $\vdash (\forall x)(\exists y)(y > x)$

is provable. In this sense it is possible to say that the class of positive integers is actually infinite. But from (1) it does not follow that

(2) $\vdash (\forall x)\mathrm{E}(y{\downarrow}(y > x)),$
(3) $\vdash (\forall x)\mathrm{M}(y{\downarrow}(y > x)),$

in other words, it does not follow that the class of positive integers is existentially and, correspondingly, potentially infinite. In order to obtain

(2) and (3), we have to assume, respectively

(4) $\vdash (\exists a)w \rightarrow \mathrm{E}(a{\downarrow} w)$,

or

(5) $\vdash (\exists a)w \rightarrow \mathrm{M}(a{\downarrow} w)$,

or to adopt (2) or (3) themselves. Since (3) follows from (2), and (5) follows from (4), the variant with assumption (2) or (4) is stronger.

24. A Remark on a Paradox with the Number Terms

Let a be an expression 'The number of planets in the solar system'. It is known that $a = 9$. According to the rules of arithmetic and logic, we have $\vdash (9 > 7)$ and 'It is necessary that $9 > 7$'. Substituting a in place of 9, we obtain; 'It is necessary that $a > 7$'. And this does not agree with intuition. The paradox results from the illegitimate use of a as a number-term. In reality the expression a has a double meaning. It can be explicated as 'A class of planets in the solar system'. But then the expression 'A class of planets in the solar system equals 9' is meaningless, since a number can equal only a number, and the expression 'A class of planets in the solar system' is not a number. It can also be explicated as 'A number which equals the cardinality of the class of planets in the solar system'. But then, there is nothing paradoxical in the expression 'It is necessary that the number which is equal to the cardinality of the class of planets in the solar system is greater than seven'.

CHAPTER FOUR

LOGICAL PHYSICS

1. On the Method of Presentation of Logic

As any other science, logic is immersed in a certain cultural and linguistic medium without which it cannot exist as a science. This fact manifests itself, in particular, in that in logic one constantly has to use certain intuitively obvious and easily understandable considerations, to make use of simple examples and explanations in ordinary language. While all this is outside of the sphere of logical laws, without this material the latter cannot be formulated and understood. And it is not a retreat from the methods of logic. On the contrary, it is a necessary element of the methodology of every science if the latter is interpreted as a whole, and not as a collection of its individual problems that may, to a certain extent, acquire an individual existence and produce an impression of independence of the above medium.

As the presentation of logic becomes more and more detailed, the need for such a common medium increases. One of the reasons for this is that logic comes closer to the surface of linguistic practice, where one has to take into account a considerable number of circumstances that violate the purity of logical norms; another reason is that logic deals with linguistic expressions that by their very nature represent a product of numerous and varied sources and function in a complex system of linguistic connections. Of the elementary purity of logical constructions which can be seen, for instance, at the level of the propositional calculus, very little is left here. In this situation, the complexities associated with deduction are replaced by those related to the necessity of coordinating a great number of various analytical factors. Because of this, the method of presentation of logic changes. At present, a presentation based on supplementary addition of new definitions and assertions, in no way affecting the previous discussion, is already impossible. A situation arises now when we have to consider at once a whole combination of problems, conceptions and statements. But it is impossible to speak about everything at once. A consistent analysis of one kind of problem implies abstraction from the others, acceptance as given of a whole number of terms and statements which in further discussion become a subject of special study. As a result of this, the presentation of material no longer corresponds with the notion of the presentation of logic formed on the basis of observation of the logic of propositions and predicates in contemporary courses of logic.

In what follows we shall from the very beginning have to make use of expressions which later will become a subject of our analysis. There is no vicious circle here at all, since we shall mean the explication of linguistic expressions that is necessarily accompanied by their 'doubling'.

2. Empirical Individuals

We shall not give the complete definition of the expression 'empirical individual' immediately. Only its partial definition, sufficient for the purposes of our presentation, will be given; it will be subject to additions as needed. The question of its complete definition will be treated again later, in connection with another problem. Partial definition is appropriate here, since we are going to consider consequences of the fact that the given objects are empirical; this consideration will be based on the available parts of the definition.

Empirical individuals possess spatial dimensions. This statement represents a part of the implicit definition of the expression 'empirical individual'. Since the meaning of the linguistic expressions recording the volumes and areas of empirical individuals reduces logically to the meaning of the expression 'length', in what follows, when considering the spatial dimensions of individuals, we shall limit ourselves to considering their length. The expression 'a length a with respect to α' will have the following symbolic form:

$$l^s\{a,\alpha\}.$$

The above statement can be written as

$$\Phi^1. \qquad \vdash \mathrm{A} \rightarrow (l^s\{a,\alpha\} > 0),$$

where A is the statement 'a is an empirical individual with respect to α'.

It follows from Φ^1 that if $l^s\{a,\alpha\} = 0$, in other words, if a has no spatial dimensions with respect to α, then a is not an empirical individual with respect to α.

For sake of argument, it is more convenient to use a different formulation

$$\Phi^1. \qquad \vdash \mathrm{B} \rightarrow (\forall\alpha)(l^s\{a,\alpha\} > 0),$$

where B is the statement 'a is an empirical individual', and α is a variable for the methods of establishing the spatial order.

3. Variation

Assume that $\downarrow x$ is a state of an empirical object a at a certain time, and $\downarrow y$ is a state of the same individual at a different time after that, i.e., $\downarrow y$ supersedes $\downarrow x$ in the temporal order. Assume also that $\vdash \sim(x \wedge y)$, in other words, that the states $\downarrow x$ and $\downarrow y$ exclude each other. We shall say that a transformation of the state $\downarrow x$ into the state $\downarrow y$ has taken place (or that $\downarrow x$ has transformed into $\downarrow y$) or that an $\downarrow x$ has transformed into an $\downarrow y$. This will be written symbolically as

$$\downarrow x \Rightarrow \downarrow y \quad \text{and} \quad a\downarrow x \Rightarrow a\downarrow y.$$

The special cases of an individual's transformation from one state into another, and especially cases of variation are as follows:

(1) $\downarrow \neg(\mathrm{E}(a) \Rightarrow \downarrow \mathrm{E}(a)$ – creation of a;
(2) $\downarrow \sim x \Rightarrow \downarrow x$ – creation of $\downarrow x$;
(3) $\downarrow \mathrm{E}(a) \Rightarrow \downarrow \neg \mathrm{E}(a)$ – annihilation of a;
(4) $\downarrow x \Rightarrow \downarrow \sim x$ – annihilation of x;
(5) $\downarrow \neg \mathrm{P}(a) \Rightarrow \downarrow \neg P(a)$ – loss of an attribute by an individual a;
(6) $\downarrow \neg \mathrm{P}(a) \Rightarrow \downarrow \mathrm{P}(a)$ – acquisition of an attribute by an individual a;
(7) $\downarrow \mathrm{P}\alpha(a) \Rightarrow \downarrow \mathrm{P}\beta(a)$, where $\alpha > \beta$ – decrease of a with respect to an attribute P;
(8) $\downarrow \mathrm{P}\alpha(a) \Rightarrow \downarrow \mathrm{P}\beta(a)$, where $\alpha < \beta$ – increase of a with respect to an attribute P.

Depending on the types of attributes with respect to which the variations of objects occur, special predicates of variation ('grew up', 'expanded', 'contracted', etc.) are introduced. This serves as a foundation for introduction of other predicates of variation, according to the rules for definitions. But their logical basis is formed by the predicate $\Rightarrow$. Moreover, all logical properties of the derived terminology, developed for recording variations, may be reduced to it.

Therefore, when we speak of variation and transformation, we imply the presence of a certain empirical object and a change of its states with time.

Transformation of one object a into another b (for instance, a pupa transforms into a butterfly) usually means transformation of a state of a certain object, denoted by the term a, into another state of the same object, denoted by the term b.

In order to call a certain set of variations, ordered in space and time, also a variation, it is necessary to introduce a term a by which one can denote a given cluster of empirical objects at different moments of time, regardless of the variations that occur, and to represent the results of the latter as different states of a.

A special case of variation is provided by change of place in space (translation or motion). We will discuss this case separately below.

Variations take place in certain spatial domains. The varying objects are characterized by spatial dimensions. But in the case of variation, one cannot introduce the spatial dimensions as such, without making certain stipulations. This is a consequence of the fact that spatial predicates are used for a different kind of empirical objects (we shall come back to this later). Variations are characterized by temporal length or duration.

In the same way as one can introduce certain spatial terms and a concept of length on the basis of observation of some empirical bodies, observation of certain variations allows to present the concept of time as well as certain elementary terminology related to it. There is also a historically conditional and temporally transient quantity characteristic of the duration of a variation, necessary for the variation to be detected.

The term 'the duration of variation of a with respect to α' will also be a symbolic form

$$l^t\{a,\alpha\}.$$

Every empirical variation takes place in real time, i.e., the duration of any empirical variation is greater than zero. This constitutes a part of the very definition of the expression 'empirical variation'.

Suppose that A is the statement 'a is an empirical variation with respect to α'. The above statement can be written as follows:

Φ^2. $\quad \vdash \mathrm{A} \rightarrow (l^t\{a,\alpha\} > 0)$.

From Φ^2 one obtains: if $l^t\{a,\alpha\} = 0$, in other words, if no time is spent on variation a, then a is not an empirical variation with respect to α.

Suppose that B is the statement 'a is an empirical variation', and α is a variable for the methods of establishing the temporal order. Then the above statement can be given a different form:

Φ^2. $\quad \vdash \mathrm{B} \rightarrow (\forall\alpha)(l^t\{a,\alpha\} > 0)$.

Applying the rules of the logic of relations to (Φ^1), we obtain:

$$\vdash \mathrm{A} \rightarrow \sim (l^s\{a, \alpha\} < 0),$$
$$\vdash \mathrm{B} \rightarrow (\forall \alpha) \sim (l^s\{a, \alpha\} < 0),$$

and from Φ^2 it follows that

$$\vdash \mathrm{A} \rightarrow \sim (l^t\{a, \alpha\} < 0),$$
$$\vdash \mathrm{B} \rightarrow (\forall \alpha) \sim (l^t\{a, \alpha\} < 0),$$

i.e., there are no empirical objects for negative lengths and durations.

4. The Transition State

Let t^1 be a moment of time for recording a state $\downarrow x$, and t^2 be that of recording a state $\downarrow y$, and assume that $\downarrow x \Rightarrow \downarrow y$. As is known from experience, there are cases when there is a time interval t^3 between t^1 and t^2, such that $\sim x \wedge \sim y$. A state $\downarrow(\sim x \wedge \sim y)$ of an individual will be called a transition state. If x is either P(a) or $\neg$P(a), and y is, respectivley, either $\neg$P(a) or P(a), then $\downarrow(\sim x \wedge \sim y)$ is $\downarrow$?P(a). Similarly, $\downarrow(\sim x \wedge \sim y)$ is $\downarrow$?E(a) when one of x and y is E(a), and the other is $\neg$E(a).

5. Space and Time

Two kinds of problems related to space and time should be distinguished: (1) terminological problems i.e., those associated with establishing the significance of such terms as 'space', 'time', 'farther', 'closer', 'earlier', etc.; (2) the problems of measurement (or physical problems), i.e., those associated with determining the spatial and temporal characteristics of objects. One can see the difference between these two aspects from the following example. If a certain person A says that for a railroad train passenger Irkutsk is closer to Moscow than Novosibirsk, for us this means an error not in the usage of the term 'closer', but in establishing the spatial order of Irkutsk and Novosibirsk for a railroad passenger with respect to Moscow. And if A agrees with the fact that Novosibirsk is closer than Irkutsk, she will use the term 'closer' in the same sense (in the same manner) as in the first case.

The division of labor in science has transformed the problems of investigating spatial and temporal properties and relations of objects into those of physics. However, physical theories of space and time do not supply definitions of spatial and temporal terminology in the sense that they do not introduce this terminology for the first time. It is adopted as given by them, and the task of the physical theories is to establish the methods of determining the spatial and temporal relations between objects for various difficult cases (as compared to everyday experience), in particular for systems moving in different ways, for remote events, in situations when one has to take into account the velocity of signal propagation, mutual motion of events and observers, and so on. It is assumed, in an explicit or implicit way, that there are certain basic cases for which establishing the spatial and temporal relations is not difficult and which serve as an example for introducing the corresponding terminology into our language. For instance, the number of basic cases includes those for which the sequence of observations of an event is regarded as coinciding exactly with that of their occurrence in time. The spatial and temporal relations for other cases are established through application of the rules of logic, mathematics, and physics to those data that are obtained from observation of the basic cases. One of the conditions and consequences of physical theories is exemplified by the explication of spatial and temporal terminology for special cases of relations between objects. The following interesting phenomenon may be observed here.

In the case of the explication just mentioned there is no introduction of new terms that take into account the peculiar features of the cases under consideration; instead, we employ the same general terminology that has been developed on the basis of conventional observations of spatial and temporal relations between objects. The result of this is implicit superposition of at least two different sets of terms which leads to well-known paradoxical consequences.

On the other hand, there was a transfer of terminology into the domain of space and time, which has been formulated for empirical objects, i.e., space and time actually came to be regarded as a sort of empirical objects, just like others. In particular, scientists introduced the notions of variation of space and time, of the course or motion of time, of space contraction, of time slowing down or accelerating, of the reversed course of time, etc. And these expressions are used not figuratively, but literally. As a result, the terminological situation in the literature, related in one or another way to space and time, has become so chaotic that a logical analysis of the most fundamental aspects of terminology would obviously be of use here.

It follows from the attempts to introduce spatial and temporal terms by means of defining them through others, that we require terms that have no advantages from the point of view of elementary obviousness as compared to spatial and temporal terms. Besides, the group of these includes terms that are introduced not through definitions but by other means. And as far as these terms are concerned, one can speak of their explication only.

The perceived form of the explicating terms may be identical to that of the explicated ones. This is exactly the situation in our case. There is no danger here whatsoever, if one keeps in mind the difference in roles of the same words.

6. Spatial and Temporal Relations

The simplest spatial and temporal terms are those from the set of values of the signs $>\alpha$, $<\alpha$ and $=\alpha$. Their relations to each other are of the same type as the relation between these signs in the general case. As for their special features, the following point has to be clarified: what is to supplement logical considerations when one or another triplet of concrete terms of order is chosen (e.g., 'earlier – later – simultaneously', 'closer – further – at the same distance', etc.)?

In general, with respect to all the simplest spatial and temporal terms of this kind the following is valid: the investigator has at her disposal special devices (innate abilities, if this is a human; special technological means if it is an artificial intelligence installation) allowing her to record spatial and temporal relations, and after a sufficiently large number of repetitions she begins to use the corresponding terms in an appropriate way (in the same manner as people learn many words of a language, denoting things, events, deeds, etc.). Among the means mentioned above are a turn of the eyes, a turn of the body, a change in perception, and so on. All these represent pre-logical operations.

The difference between objects, appearing in spatial and temporal order, is important. In the case of spatial order the set of objects is that of empirical bodies (perceived, i.e., seen, heard, etc.). In the case of temporal relations, the objects are perceived changes (the occurrences of events, 'explosions', 'perturbations', and so on).

If one ignores the lengths and durations of certain objects and variations and takes into account their order with respect to each other and the intervals between them, these objects and variations take on the role of special marks or signs of space and time, respectively. In such cases we shall introduce the notions of points in space and moments of time. The expressions 'at a given point in space' and 'at a given moment' should not be understood in the sense that the corresponding bodies and variations have no length and duration. These expressions account just for the place of objects in spatial or temporal order.

The set of terms, discussed above, represents a set of predicates. By making use of it, we can produce certain statements concerning spatial and temporal properties and relations of empirical individuals. But these terms cannot be employed as subjects of statements; in other words, it is still impossible to say something about space and time as special objects.

In order to introduce terms that denote space and time as special objects, the following operations of term introduction should be performed.

Assume that empirical individuals $a^1,\ldots,a^n$ form a spatial structure A. There are at least three different kinds of meaning possible for the term 'a given space A'.

With its first meaning the term 'a given space A' is introduced as follows. We consider a spatial structure A and ignore the presence of all individuals inside A, as well as the fact that it is formed exactly of $a^1,\ldots,a^n$. This abstraction is given an intuitively obvious appearance of assumptions that agree well with everyday experience: the individuals inside A can be taken out of it (the limiting case – an empty space without individuals), arbitrary individuals of suitable dimensions can be placed in A, one can replace individuals $a^1,\ldots,a^n$ by any others which would be appropriate for fixing the boundaries of A. Finally, by the term 'a given space A' we shall mean only what is situated between certain individuals the distribution of the latter being similar to that of $a^1,\ldots,a^n$ in A after all the individuals have been excluded from A. The individual character of a given space is determined by the position of A.

Thus, the above meaning of the term 'a given space A' implies what is left when only the position of A is preserved, and all the individuals are taken out of A (including its boundary points $a^1,\ldots,a^n$). Thus, it would be more convenient to read the term 'a given space A' as 'a space, occupied by A'.

The following is of consequence here: whether the above operation is or is not possible to be carried out, it in no way affects the definition of the term 'a given space A' or its legitimacy.

The formation of the second meaning of the term denoting a given space A requires the same operation to be performed as in the first case, excluding the assumption about the objects $a^1,\ldots,a^n$: they are regarded as boundary points of a given space. The third meaning of the term 'a given space A' is associated with an assumption to the effect that it is formed by a spatial structure A together with certain objects inside it. Different situations are possible here; in one of them, a given space A includes all objects inside A, and in this case a given space A represents a piece of matter located inside the spatial structure A; in other situations it includes only certain special kinds of 'fine' matter (in particular, those that cannot in principle be taken out of A under any circumstances).

The terms of a given time exhibit a similar pattern. We consider a temporal structure B formed by events $b^1, \ldots, b^n$ and formulate assumptions with respect to events $b^1, \ldots, b^n$ inside B in the same manner as those for terms of a given space. We would like only to remark that the assumptions concerning time, which are similar to those concerning space, can be interpreted in a visually obvious manner from the point of view of everyday experience. The point is that a scientist fixes a certain spatial structure at a certain time, and a temporal structure in a certain spatial domain. On the other hand, in everyday experience one apparently may come across domains of space where nothing happens ('everything is quiet'). Besides, the investigator may, within the frame of a specified temporal structure, provoke suitable events by herself or hinder their occurrence. At any rate, she also may pay no attention to them. The third meaning of the term 'a given time B' is formed by the world with all (or selected) events in a spatial structure B.

The terms 'a given space' and 'a given time' represent (like the term 'structure') a generalization of terms denoting concrete spaces and times. Certainly, there is a complex of diversified terms in real languages, such as 'interval', 'location', 'segment', 'spatial domain', etc., but all of them represent special cases, in the same way as the terms 'binary structure', 'ternary structure', etc. represent the species terms for the term 'structure'.

But ordinarily, whenever the terms 'space' and 'time' are used independently (as different from the term 'structure' and related terms, denoting concrete structures), rather than implying arbitrary concrete spaces and times ('intervals' and 'domains' of space and time), one has in mind their being unified into a whole. In order to introduce these terms (let us say, 'space as a whole', 'time as a whole'), it is necessary to assume the existence of spatial and temporal structures that include arbitrary given spatial and correspondingly, temporal structures (in the sense discussed above), and to satisfy the same requirements with respect to this hypothetical structure as those with respect to given spatial and temporal structures. However, in the course of this procedure the difference between the first and the second meanings of terms is lost. What remains is only the difference between the first and the third ones: (1) space is a container of all things, time is pure duration; (2) space is the world of all (or selected) things, time is the world of all (or selected) events.

Thus, an explicit diversity in the meanings of terms is revealed. And at the same time the future of many statements with spatial and temporal terminology depends entirely on what definitions have been adopted (explicitly or implicitly) for it. And no scientific achievements are capable of alleviating this dependence. If, for instance, one considers a statement to the effect that in the vicinity of such and such star space is curved, this statement means absolutely nothing until one specifies the meaning of the term 'space'. If it is the first meaning, the statement is absurd, since only a certain series of objects can be curved with respect to another series, and from the point of view of the first concept of space all this is excluded. Therefore, the term 'space' is used here with its third meaning. But then there is nothing surprising in this.

We would like to remark concerning space and time as a whole (i.e., the unification of all spaces and times). In order to unify all spatial (temporal) structures into a single spatial (temporal) one, the same class of methods of establishing spatial (temporal) order is required for all structures. And this is actually unthinkable. On the other hand, if one has in mind the class of all methods of establishing order (and there will be indeed only one such class), it will contain methods that are mutually exclusive and therefore it will be logically impossible. Thus, all kinds of discussion concerning space and time as a whole (unification of all spaces and times) always remain purely hypothetical.

Moreover, an attempt to give the above abstractions logically explicit form results in the following interesting fact.

An abstraction performed in order to produce the first meaning of the term 'a given space A' can be explicitly expressed follows: If one takes out (excludes) from the spatial structure A all the individuals located inside A, as well as the individuals $a^1, \ldots, a^n$, leaving all the rest unchanged, then what is left of A will be called the space A (or, the space occupied by A). The right to use the term 'the space A' in this statement depends on the fulfilling of such a condition, which is unrealizable. And therefore, consistent with the properties of the operator 'if, then', this right cannot be applied. The substitution of the expressions 'if something would have happened, then', 'if it were possible to, then' in place of 'if, then' leads to a still greater degree of confusion. Hence the absolute logical inconsistency of the above schemes of formation of spatial terms. The situation with time is similar.

We see as logically correct only the following method. Logical explication can be applied to terms denoting the spatial and temporal order of individuals, to such terms as 'spatial structure', 'temporal structure', and to the set of terms, related to these, which we have discussed to some extent. As far as the words 'space' and 'time' themselves are concerned, they may be defined precisely only as parts of complex expressions (if it is needed, of course). For example, in the following manner: 'An individual a is in a space A' $\equiv$ Df. 'An individual a is inside a spatial structure A'; 'Space is curved in the domain α' $\equiv$ Df. x, where x represents a description of data from observation or experiment with respect to the behavior or position of certain individuals in the domain α. However, in all such cases it is actually the expressions containing the words 'space' and 'time', and not these words themselves as separate independent terms, that are being defined.

7. The Time and Space of Existence of an Empirical Individual

Such statements as 'a exists at a time t' will be given a symbolic form

$$Et(a),$$

where t is a certain notation for time (temporal interval or moment), and Et is a predicate 'exists at a time t'.

Let us further assume that temporal intervals (times) and moments are ordered with respect to a certain method of establishing order. In order to simplify the record, in this section we shall use the symbols $t^1 > t^2, t^1 < t^2$ and $t^1 = t^2$ as denoting, respectively, the following expressions: 't^1 is after (later than) t^2', 't^1 is before (earlier than) t^2', 't^1 and t^2 coincide (represent the same time)'.

A statement 'The time of existence of an empirical individual is localized and continuous' (X) should be regarded as a part of an implicit definition of the expression 'an empirical individual'. This statement has the following meaning. Let t^1 be a time when a appears (it did not exist before), and t^2, a time when a either disappears or is destroyed (it does not exist after this). At any time t^i such that $t^i < t^1$ a does not exist (statement X^1). At any time t^i such that $t^i > t^2$ a does not exist (statement X^2). At any time t^i such that $t^2 > t^i$ and $t^i > t^1$ a exists (statement X^3). In other words, a does not exist at any time before t^1 (X^1), a does not exist at any time after

t^2 (X^2), and if a exists at t^1 and at t^2, it does exist at any time in the interval between t^1 and t^2. It would be even appropriate to say with respect to the time after t^2 that a is impossible at any time after t^2. The statement X represents just an abbreviation for $X^1 \wedge X^2 \wedge X^3$, i.e., $X \equiv \text{Df. } X^1 \wedge X^2 \wedge X^3$. And the latter is accepted as an axiom representing a part of an implicit definition of an empirical individual.

Assertions X^1, X^2 and X^3 may be obtained as consequences of other axioms, namely, the following ones:

Φ^3. If a exists at t^1 and does not exist at t^2 before t^1, it does not exist at any time before t^2.

Φ^4. If a exists at t^1 and does not exist at t^2 after t^1, it does not exist at any time after t^2.

It follows from Φ^3 and Φ^4, that if a exists at t^1 and at t^2, and if $t^1 > \alpha t^2$ or $t^2 > \alpha t^1$, then a exists at any time t^3 between t^1 and t^2. The validity of this assertion is obvious: if a does not exist at t^3, then a does not exist at t^1, if $t^3 > t^1$ or a does not exist at t^2 if $t^3 > t^2$; also, a does not exist at t^1, if $t^3 < t^1$ or at t^2 if $t^3 < t^2$.

Let α, β and γ be the variables for terms of time. Assertions Φ^3 and Φ^4 may be also written as

Φ^3. $\vdash (\mathrm{E}\alpha(a) \wedge (\exists\beta)((\beta < \alpha) \wedge \sim \mathrm{E}\beta(a)) \rightarrow$
$\rightarrow (\forall\gamma)((\gamma < \beta) \rightarrow \sim \mathrm{E}\gamma(a)),$
$\vdash (\mathrm{E}\alpha(a) \wedge (\exists\beta)((\beta > \alpha) \wedge \sim \mathrm{E}\beta(a)) \rightarrow$
$\rightarrow (\forall\gamma)((\gamma > \beta) \rightarrow \sim \mathrm{E}\gamma(a)).$

From Φ^3 and Φ^4 one can obtain the assertion of continuity for the time of existence of an empirical individual:

T. $\vdash \mathrm{E}\alpha(a) \wedge \mathrm{E}\gamma(a) \rightarrow (\forall\beta)(((\beta < \alpha) \wedge (\gamma < \beta)) \vee$
$\vee ((\beta > \alpha) \wedge (\gamma > \beta)) \rightarrow \mathrm{E}\beta(a)).$

The following is a more or less complete presentation of the proof of T. For the sake of brevity, instead of formulas Φ^3 and Φ^4 we shall write their numbers Φ^3 and Φ^4. If a certain theorem constitutes a part of the theorem following after it in the presentation of the proof, the proof will contain just its number. If N is a number of a certain theorem that has the form $N^1 \rightarrow N^2$, then the symbol $(N \rightarrow)$ will stand for its antecedent N^1, and the symbol $(\rightarrow N)$ – for its consequent N^2.

The proof of T.

(1) $(\rightarrow \Phi^3) \vdash ((\gamma < \beta) \rightarrow \sim \mathrm{E}\gamma(a)).$

(2) $\vdash((\rightarrow \Phi^3) \rightarrow ((\gamma < \beta) \rightarrow \sim \mathrm{E}\gamma(a)))$.
(3) $\vdash \Phi^3 \wedge 2$.
(4) $(\Phi^3 \wedge 2) \vdash ((\Phi^3 \rightarrow) \rightarrow ((\gamma < \beta) \rightarrow \sim \mathrm{E}\gamma(a)))$.
(5) $\vdash((\Phi^3 \rightarrow) \rightarrow ((\gamma < \beta) \rightarrow \sim \mathrm{E}\gamma(a)))$.
(6) $\mathrm{E}\alpha(a) \wedge ((\beta < \alpha) \wedge \sim \mathrm{E}\beta(a)) \vdash (\beta < \alpha) \wedge \sim \mathrm{E}\beta(a)$.
(7) $(\beta < \alpha) \wedge \sim \mathrm{E}\beta(a) \vdash (\exists \beta)((\beta < \alpha) \wedge \sim \mathrm{E}\beta(a))$.
(8) $\mathrm{E}\alpha(a) \wedge ((\beta < \alpha) \wedge \sim \mathrm{E}\beta(a)) \vdash (\exists \beta)((\beta < \alpha) \wedge \sim \mathrm{E}\beta(a))$.
(9) $\mathrm{E}\alpha(a) \wedge ((\beta < \alpha) \wedge \sim \mathrm{E}\beta(a)) \vdash \mathrm{E}\alpha(a)$.
(10) $\mathrm{E}\alpha(a) \wedge ((\beta < \alpha) \wedge \sim \mathrm{E}\beta(a)) \vdash (\Phi^3 \rightarrow)$.
(11) $\mathrm{E}\alpha(a) \wedge (\beta < \alpha) \wedge \sim \mathrm{E}\beta(a) \vdash \mathrm{E}\alpha(a) \wedge ((\beta < \alpha) \wedge \sim \mathrm{E}\beta(a))$.
(12) $\mathrm{E}\alpha(a) \wedge (\beta < \alpha) \wedge \sim \mathrm{E}\beta(a) \vdash (\Phi^3 \rightarrow)$.
(13) $\vdash((\mathrm{E}\alpha(a) \wedge (\beta < \alpha) \wedge \sim \mathrm{E}\beta(a)) \rightarrow (\Phi^3 \rightarrow))$.
(14) $\vdash 13 \wedge 5$.
(15) $13 \wedge 5 \vdash (\mathrm{E}\alpha(a) \wedge (\beta < \alpha) \wedge \sim \mathrm{E}\beta(a)) \rightarrow ((\gamma < \beta) \rightarrow \sim \mathrm{E}\gamma(a))$.
(16) $\vdash \mathrm{E}\alpha(a) \wedge (\beta < \alpha) \wedge \sim \mathrm{E}\beta(a) \rightarrow ((\gamma < \beta) \rightarrow \sim \mathrm{E}\gamma(a))$.
(17) $\vdash \mathrm{E}\alpha(a) \wedge (\beta < \alpha) \wedge \sim \mathrm{E}\beta(a) \wedge (\gamma < \beta) \rightarrow \sim \mathrm{E}\gamma(a)$.
(18) $\vdash \mathrm{E}\alpha(a) \wedge \sim \mathrm{E}\beta(a) \wedge ((\beta < \alpha) \wedge (\gamma < \beta)) \rightarrow \sim \mathrm{E}\gamma(a)$.

A similar method can be applied to the proof of Formula 19:

(19) $\vdash \mathrm{E}\alpha(a) \wedge \sim \mathrm{E}\beta(a) \wedge ((\beta > \alpha) \wedge (\gamma > \beta)) \rightarrow \sim \mathrm{E}\gamma(a)$.
(20) $\vdash(18 \rightarrow) \rightarrow \mathrm{E}\alpha(a)(\beta < \alpha) \wedge (\gamma < \beta) \wedge \sim \mathrm{E}\beta(a)$.
(21) $\vdash(\rightarrow 20) \rightarrow \sim \mathrm{E}\gamma(a)$.
(22) $\vdash \mathrm{E}\alpha(a) \wedge (\beta < \alpha) \wedge (\gamma < \beta) \wedge \mathrm{E}\gamma(a) \rightarrow \mathrm{E}\beta(a)$.
(23) $\vdash \mathrm{E}\alpha(a) \wedge \mathrm{E}\gamma(a) \wedge ((\beta < \alpha) \wedge (\gamma < \beta)) \rightarrow \mathrm{E}\beta(a)$.

Formula 24 is obtained in a similar way:

(24) $\vdash \mathrm{E}\alpha(a) \wedge \mathrm{E}\gamma(a) \wedge ((\beta > \alpha) \wedge (\gamma > \beta)) \rightarrow \mathrm{E}\beta(a)$.
(25) $\vdash(23 \vee 24) \rightarrow \mathrm{E}\beta(a)$.
(26) $\vdash \mathrm{E}\alpha(a) \wedge \mathrm{E}\gamma(a) \wedge (((\beta < \alpha) \wedge (\gamma < \beta)) \vee ((\beta > \alpha) \wedge$
$\wedge (\gamma > \beta))) \rightarrow \mathrm{E}\beta(a)$.
(27) $\vdash \mathrm{E}\alpha(a) \wedge \mathrm{E}\gamma(a) \rightarrow ((((\beta < \alpha) \wedge (\gamma < \beta)) \vee ((\beta > \alpha) \wedge$
$\wedge (\gamma > \beta)) \rightarrow \mathrm{E}\beta(a))$.
(28) $\vdash(\forall \beta)(27)$.
(29) $\vdash 28 \rightarrow ((27 \rightarrow) \rightarrow (\forall \beta)(\rightarrow 27))$.
(30) $\vdash(27 \rightarrow) \rightarrow (\forall \beta)(\rightarrow 27)$.

Formula 30 is T.

Assertions 18 and 19 are equivalent to Φ^3 and Φ^4 in the sense that if 18 and 19 are accepted as axioms, then Φ^3 and Φ^4 will be theorems, deducible from them.

If one takes into account the difference between negations $\sim$ and $\neg$, then axioms Φ^3 and Φ^4 should be replaced by the following four statements:

(1) $\vdash (\Phi^3(\sim E\beta(a)/\neg E\beta(a)))(\sim E\gamma(a)/\neg E\gamma(a))$.
(2) $\vdash (\Phi^3(\sim E\beta(a)/?E\beta(a)))(\sim E\gamma(a)/?E\gamma(a))$.
(3) $\vdash (\Phi^4(\sim E\beta(a)/\neg E\beta(a)))(\sim E\gamma(a)/\neg E\gamma(a))$.
(4) $\vdash (\Phi^4(\sim E\beta(a)/\neg ?E\beta(a)))(\sim E\gamma(a)/?E\gamma(a))$.

A part of the definition of an empirical individual is represented by assertions:

Φ^5. $\vdash E\alpha(a) \wedge (\exists\beta)((\beta < q\alpha) \wedge \sim E\beta(a) \rightarrow (\forall\gamma)((\gamma < q\beta) \rightarrow \sim E\gamma(a))$,
Φ^6. $\vdash E\alpha(a) \wedge (\exists\beta)((\beta > q\alpha) \wedge \sim E\beta(a) \rightarrow (\forall\gamma)((\gamma > q\beta) \rightarrow \sim E\gamma(a))$,

where α, β, γ are variables for the terms of space. An assertion T^2, which is similar to T^1 and reads as a statement of continuity for the space of existence of an empirical individual or as that of its being unique in space.

It seems as if statement T^2 contradicts those cases when an empirical individual consists of parts separated in space. Denote the individual by A, and its parts by B and C. In the interval between them there is no B and there is no C. But we cannot say that A exists at the location of B or C, and A does not exist in the interval between B and C. This interval also represents a space, occupied by A.

The duration of existence of an empirical individual is greater than zero, i.e.,

Φ^7. $\vdash (\exists\alpha)E\alpha(a) \rightarrow (\exists\beta)(E\beta)(a) \wedge (1\beta > 0))$,

where α and β are the variables for intervals of time.

8. The Existence of Space and Time

The construction of spatial and temporal terminology does not end with the introduction of the appropriate subjects. If those are introduced, it does not mean that we can obtain rational statements combining them with arbitrary predicates. The conditions for predication with space and time still have to be defined for a whole series of predicates. Consider, for instance, a statement about the infinity of space. We can see its meaning, in particular if it is interpreted as a replacement for the following assertion: there is a method of establishing an order with respect to which one orders the spatial structures; and this ordered series is infinite. Otherwise the above statement has no meaning.

Take the existence predicate. What is the meaning of statements concerning the existence of space and time? Leaving aside the differences in terminology considered above, we shall concentrate on more general problems: (1) the question of the existence of a given space and time and that of space and time in general; (2) the question of the existence of spatial and temporal structures.

As far as the first issue is concerned, there are no fundamental difficulties: the question of the existence of any arbitrary space (and time) coincides with that of the existence of space (time) in general; both of them may be reduced to the question of the existence of a certain given space (and time), in the same way as the question of the existence of table in general may be reduced to the question of the existence of a certain specific table.

The existence of structures depends on the existence of the order-relations, and the question of the existence of the latter reduces to the question of the existence of objects: a relation aRb exists if and only if both a and b exist, the relation between them being exactly as stated in the assertion aRb. Other definitions of existence represent just more complex versions of this scheme.

Since spatial structures are structures of empirical objects, it follows from the very definition of the existence predicate for these structures that space without empirical objects does not exist. This fact remains latent, however, since not only in the language of everyday life but in science as well, implicit definitions are frequently used, and the thesis of the impossibility of the existence of space without objects is interpreted as a result of observations or as a postulate. The situation with time is quite similar: time does not exist without empirical variations.

From the investigator's point of view, a given space A exists if and only if a given spatial structure A does exist (or if certain objects forming the spatial structure A [this term is used with its first meaning] exist). And the existence of the latter for the investigator depends entirely on the existence of objects $a^1, \ldots, a^n$ (or of some objects at the appropriate locations), and their order is such as specified in the definition of A. In general, space exists for the investigator if and only if she recognizes the existence of a certain given space. It is important to note that here we assume the existence of objects that form a given spatial structure at a certain given time (it is implied that things occur 'simultaneously').

The definitions of existence for time may be constructed in a similar manner, but one must keep in mind that their statements deal with variations of things rather than with things that are empirically stable.

And in this case it is the fact that variations occur at different times that is of importance. The investigator observes variations, forming a given temporal structure; therefore, they exist for her at different times while the temporal structure itself exists for the investigator exactly when it is recorded by her. Thus, from the investigator's point of view, it makes sense to speak not of the time of existence of a given time, but only of a given spatial domain in which the investigator observes the variations which form a given temporal structure. We may say that these variations are 'single-spatial'.

The time and location of the existence of time and, respectively, of space, are specified by the following assertions:

Φ^8. $\vdash \mathrm{E}t(t)$,
$\vdash \mathrm{E}s(s)$,
$\vdash (t^1 > t^2) \rightarrow \neg \mathrm{E}t^1(t^2) \wedge \neg \mathrm{E}t^2(t^1)$,
$\vdash (s^1 > \alpha s^2) \rightarrow \neg \mathrm{E}s^1(s^2) \wedge \neg \mathrm{E}s^2(s^1)$.

We shall say that an interval of time t^1 is included within an interval of time t^2 with respect to α if and only if all variations that occur within t^1 with respect to α occur within t^2 with respect to α. We shall also say that a spatial structure s^1 is included within a spatial structure s^2 with respect to α if and only if all the bodies located inside s^1 with respect to α are located inside s^2 with respect to α.

9. The Spatial and Temporal Location of an Individual

The location of individuals in space and time is determined by their place in spatial and temporal series as well as by their position within the spatial and temporal structures. In what follows, whenever the expressions 'at a time t' and 'at a location s' are used, we shall have in mind a given (arbitrary) method of establishing the position of an individual.

10. One and the Same Individual

One has to distinguish between two uses of the term 'identical' with respect to individuals:

(1) when it is implied that individuals *a* and *b* are identical in certain attributes; in particular, when *a* and *b* are identical in all attributes;

(2) when it is implied that *a* and *b* represent the same individual.

The statement of the special case, described in (1), can be written as

(P is the variable for predicates):

$$(\forall \mathrm{P})(\mathrm{P}(a) \leftrightarrow \mathrm{P}(b)).$$

In this case we may say that a and b are identical (in everything).

It seems natural to believe that if a and b are identical in everything, then a and b represent one and the same individual. However, the meaning of the expression 'the same individual' is quite different. Consider the following example: an individual a has an attribute P at a time t^1 and has no attribute Q at this time; then variation occurs for it, and at a time t^2 the individual a has no P, but it acquires Q. The individual a remains the same, but the individual a at a time t^1 and the individual a at a time t^2 are not identical in everything.

The expression 'the same individual' is defined in a completely different manner, namely, in the following way. Individuals a and b represent the same individual if and only if the assertions

(1) $(\forall\alpha)(\forall\beta)(= \beta b),$

(2) $(\forall\alpha)((\mathrm{E}\alpha(a) \leftrightarrow \mathrm{E}\alpha(b)),$

hold for them (in the above formulas α is a variable for time, and β is a variable for the methods of establishing the position of an individual in space). According to the first assertion, at any time a is identical to b in its spatial position with respect to any method of establishing spatial order. According to the second assertion, whenever one of a and b exists, the other one exists too (i.e., the time moments of their existence coincide). Applying the rule of contraposition, we obtain from the second assertion the result that whenever one of a and b does not exist the other one does not exist either.

Thus, the expression 'the same individual' is widely used in practice. As for the case when the empirical individuals a and b are identical in everything, we find it actually cannot occur since empirical individuals are subject in one way or another to constant variations (this statement is derived from experiment). Only with respect to abstract individuals can one assume that some of them do not vary with time.

The expression 'the same' was defined above for individuals. Further, it has to be defined for other types of empirical objects, for classes, clusters, structures, etc. For us its definition for structures is of special interest here.

Consider a structure A formed by individuals $a^1, \ldots, a^n$, and a structure B formed by individuals $b^1, \ldots, b^m$. In order for A and B to be regarded as the same structure, the following are necessary: (1) the number of individuals $a^1,\ldots,a^n$ should be equal to the number of individuals $b^1,\ldots,b^m$; (2) for every individual a^i there is an individual b^k such that a^i and b^k represent the same individual; and vice versa, for each b^k there is an a^i such that b^k and a^i represent the same individual. But this is not sufficient. There still remain the relations betwen individuals forming each structure. Are we or are we not to include the identity of these relations in the definition of the expression 'the same' for structures? There is nothing in the nature of things which would impose an obligation on us to make one or another decision.

Thus, at least two meanings of the expression 'the same' as applied to structures are possible:

(1) the weaker meaning, whereby it is implied that the above two points are sufficient for the definition (we shall call it 'weakly the same');

(2) the stronger meaning, where in addition to the above two points the definition includes the third one i.e., the identity of relations between corresponding pairs of individuals (this will be called 'strongly the same').

The strong version, in its turn, can be differentiated: it is possible to limit ourselves to the requirement of identity of relations at the level of order predicates, but it is also possible to specify that the corresponding intervals between individuals should coincide. Thus, as follows from Figure 8, the order relation is preserved in the sense that in the two cases 2 and 4, $b^2 > \alpha b^1$, but the interval between b^1 and b^2 has changed. Therefore, while we are dealing with the same individual in the pair a^1 and b^1 and in the pair a^2 and b^2, in order to answer the question as to whether we are dealing with the same structure or with different structures, an agreement about the order of individuals and the duration of intervals is still required. While in the weak version all structures 1–4 represent the same structure, in the strong version structures 1 and 3 are not the same. Taking the dimensions of intervals in the strong version into account

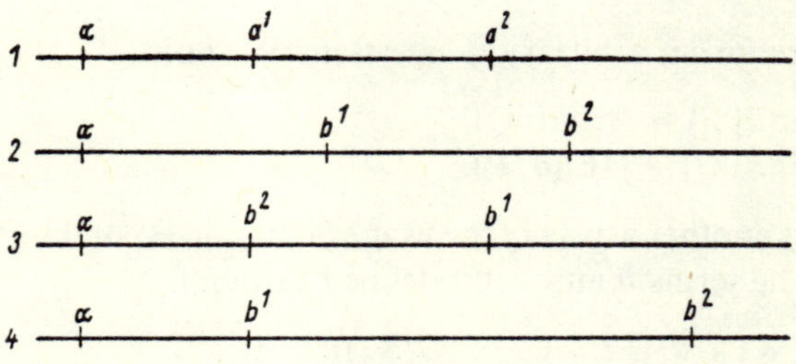

Fig.8.

results in that all structures 1–4 are pairwise not the same structure.

We would like to draw the reader's attention to the fact that the negation of the statement '*a* and *b* represent the same individual' is the statement '*a* and *b* do not represent the same individual', and *not* the statement '*a* and *b* are different'. The point is that while *a* and *b* may be different, they may represent the same individual at different times.

Consider an interesting paradox arising in connection with the usage of the expression 'the same'.

Assume that *b* is the term '*a* is a child', and *c* is the term '*a* is an adult', where *a* is an individual term (the name of a person, individualizing her). Thus, *b* and *c* represent the same individual *a* at different times. And depending on time, this individual is characterized by different attributes, in one case by the attributes of a child, in the other by the attributes of an adult. The relation between terms *a*, *b*, *c* is as follows:

$$(b \rightharpoonup a) \wedge (c \rightharpoonup a) \wedge \sim(b \rightharpoonup c) \wedge \sim(c \rightharpoonup b).$$

If *b* and *c* represent the same individual, for any time *t*

$$\mathrm{E}t(b) \leftrightarrow \mathrm{E}t(c).$$

However, a person cannot be both a child and an adult, i.e., it appears legitimate that

$$\mathrm{E}t(b) \rightarrow \neg \mathrm{E}t(c),$$
$$\mathrm{E}t(c) \rightarrow \neg \mathrm{E}t(b).$$

This paradox results from the violation of the rules of logic. The following two aspects must be distinguished here. In fact, there are such attributes of a child P and of an adult Q that for any person

$$\mathrm{P}(d) \rightarrow \neg \mathrm{Q}(d),$$
$$\mathrm{Q}(d) \rightarrow \neg \mathrm{P}(d).$$

And therefore for an arbitrary *t*, assertions W^1 hold:

$$\mathrm{E}t(d \downarrow \mathrm{P}) \rightarrow \neg \mathrm{E}t(d \downarrow \mathrm{Q}),$$
$$\mathrm{E}t(d \downarrow \mathrm{Q}) \rightarrow \neg \mathrm{E}t(d \downarrow \mathrm{P}).$$

But there is another aspect of the usage of the corresponding expressions, as follows. The terms *b* and *c* are defined through

$$b = \mathrm{Df.}\ a \downarrow (\mathrm{P}t^1(a) \wedge \neg \mathrm{Q}t^1(a)),$$
$$c = \mathrm{Df.}\ a \downarrow (\mathrm{Q}t^2(a) \wedge \neg \mathrm{P}t^2(a)),$$

where $t^2 > t^1$. And since b and c represent the same individual, we obtain an assertion W^2:

$$Et(b) \leftrightarrow Et(c),$$

which means, according to the definition of b and c, the following:

$$Et(a \downarrow (Pt^1(a) \wedge \neg Qt^1(a))) \leftrightarrow Et(a \downarrow (Qt^2(a) \wedge \neg Pt^2(a))).$$

And this formula contains nothing paradoxical. The paradox results from confusion of different aspects when both W^1 and W^2 are obtained, i.e., from their 'superposition'.

11. Change in Space and Time

Consider a spatial structure A, formed by objects $a^1, \ldots, a^n$. In order for the structure A taken at a time t^1 and the structure A taken at a different time t^2 to be regarded as the same structure, it is necessary that for each a^i we have: a^i, taken at t^1, and a^i, taken at t^2, represent the same object. But is it a necessary condition or not? Does one have to specify in the definition that all the relations between a^i at t^2 remain the same as at t^1?

If this second requirement is not specified in the definition (i.e. considered as not necessary), the spatial structures, by virtue of definitions, will be able to change (i.e., a statement 'A has changed' as well as other derived statements concerning the magnitudes of changes, their rate, etc. may be true). If, on the other hand, this requirement is regarded as necessary, then all spatial structures must remain invariable by definitions: if the relations between a^i's at t^2 have not changed as compared to those at t^1, the structure A has not changed; if, however, they have changed, we have no right to say that A at t^1 and A at t^2 represent the same structure, and have no conditions for application of the predicates of change.

The second requirement does not interfere with the fact that one kind of structure is replaced by others. And everything that may be said without adopting the second condition may be also said with it present, only in somewhat different expressions (in a different version of the language).

Incidentally, there is another interesting circumstance. An object a, considered at any time, represents the same object a. The specification of time here does not constitute a part of the term. If, on the other hand, we introduce the specification of time in the term, in other words, consider the terms 'a taken at t^1' ('a, which exists at t^1') and 'a taken at t^2' ('a, which exists at t^{2}') under the condition that one of t^1 and t^2 succeeds the other, then the objects denoted by them cannot be regarded as the same object because of the second point of the definition. They just represent elements of the same class of objects *a*. Thus, if in specification of a spatial structure A there are names of the objects forming it as discussed above, the conditions for the application of the predicates of change disappear, and all statements concerning the changes of spatial structures turn out to be unverifiable.

Now consider temporal structures. In this case, in order to be able to say 'the same spatial structure', it is required that the events forming it remain the same. Let us turn to the simplest case: a structure B consisting of two events b^1 and b^2. The time of existence of B is B itself. Assume that b^1 occurred at t^1, and b^2 at t^2. The interval between t^1 and t^2 is a time t_1. In order to observe the variation, we have to consider the same structure B at a different time t_2. The point is that both t_1 and t^1, and t_2 and t^2 represent different times. But according to the definition of existence of events the events b^1 and b^2 do not exist at t_2. If one considers certain events, similar to b^1 and b^2, but occurring at t_2, it follows from the definitions that they do not occur at t_1. Thus, in this situation no statements concerning the variations (changes) of temporal structures, regardless of the condition that is similar to the second condition for spatial structures, can be verified. In this case one cannot say that a temporal structure has changed. However, one also cannot say that it has not changed: in order to accept the fact that B has not changed, it is necessary to reproduce the same events b^1 and b^2 at t_2, which is impossible. Therefore, because of the definitions our statements concerning both the variations of time (its acceleration, slowing down, etc.) and its constancy remain unverifiable.

But can our acceptance of such statements be regarded as harmless? By no means. If we have accepted an assertion that temporal structures vary, then, because of the meaning of the corresponding terms, we will reach conclusions that contradict the definitions adopted (for instance, the conclusion that one and the same event occurs at different times). Acceptance of such statements means implicit rejection of given definitions, a change in the meaning of terminology, etc. This concerns not physical assumptions, but just careless usage of words.

As far as variations are concerned, the only appropriate thing to say with respect to temporal structures is that there may be variation of a temporal interval between different elements of classes of events, variation of the order of the elements from classes of events, etc. But this has nothing to do with the variation of temporal structures and time.

Can a situation be imagined such that at one location of the world, time passes differently (faster, more slow, in a reverse direction) as compared to another? Let alone the fact that it is in general absurd to speak of the motion of time if one uses the words with their precise meaning and considers motion as a kind of variation, there are some words appearing in this case that represent a result of comparison (for instance, the word 'faster'). Let us forget about this absurdity and formulate the problem in an acceptable way: assume that events a^1 and a^2 are elements of a class of events Ka (i.e., uniform events), while b^1 and b^2 belong to a class Kb. Assume that the measurement of time between a^1 and a^2 resulted in $a^1(\mathrm{R}^1\alpha)a^2$, and the measurement of time between b^1 and b^2 produced a result $b^1(\mathrm{R}^2\beta)b^2$. In order to compare these times, one common method is necessary for establishing the temporal relation both for the pair of events a^1 and a^2 and for the pair b^1 and b^2. Only after quantities $a^1(\mathrm{R}^3\gamma)a^2$ and $b^1(\mathrm{R}^4\gamma)b^2$ are obtained, can we express our opinion about the temporal relation between a^1 and b^1, on the one hand, and between a^2 and b^2, on the other, on the basis of their comparison. For instance, we can say that in a certain spatial domain A first occurs c, and then d, while in a spatial domain B, vice versa, the interval between c and d in A being twice that in B.

Variation of the duration of existence of individuals accompanying variation of certain conditions cannot be regarded as variation of the course of time. Thus, variation in the velocity of motion determines the duration of existence (life time) of certain microscopic particles. But to interpret this as an example of variation (acceleration or slowing down) in the course of time is as legitimate as to see variation in the course of time in the case of variation of people's life span depending on variation in food quality.

Consider, finally, the following situation. Let α and β be a pair of identical clocks. The word 'identical' is used here as a literary expression without any commitments. To put it more accurately, we must say that the clocks are uniform or belong to one class. Suppose that α is left on the Earth (A), and β is placed on a body B which leaves the Earth at a time t^1, moves in a certain way and returns to the Earth at a time t^2. Suppose that the readings of α and β turn out to be different (in α an event occurred, characterized by a quantity x, while in β it is characterized by a quantity y, x and y being non-equal).

What can one say about this fact if one takes care to observe the rule to use meaningful expressions? The observer who performs the comparison of α and β and who is in possession of a certain method of the measurement of time, in which he fixes t^1, t^2 and the interval z between these, can say the following: the readings of α and β are different: during the same time z a certain process took place in α, characterized by a quantity x, whereas in β a process of the same kind occurred, characterized by a quantity y; in one of the clocks α and β, this process proceeded more slowly (faster) than in the other one. The question about how much time has elapsed should be counterposed with another question: with respect to what method of measuring time? And only under this condition can we produce the following answers that do not contradict each other: (1) during the time z a time has elapsed which is characterized by a quantity x (a time x elapsed), with respect to α set up on A; (2) during the time z a time y has elapsed with respect to β set up on B; (3) a time z has elapsed with respect to a certain method of establishing temporal relations; (4) a time x has elapsed with respect to α on A; (5) a time y has elapsed with respect to β on B. It does not matter that (1)–(5) may pairwise coincide and differ. But it is impossible to construct meaningful expressions such as 'Different times have elapsed for A and B', 'Time passed at different rate for A and B', etc., since they imply comparison and the same time, without which this comparison is logically excluded.

Therefore, it is not our point that the assertion of slowing down or acceleration of time is wrong. The point is that in this case this assertion itself is as meaningless as its negation. Here we come across linguistic constructions which resemble statements but are not; the reason for this is that the expressions occurring in these linguistic constructions have no meaning (are not terms). And a meaningless statement can neither be proved (confirmed) nor disproved.

If for some objects a time x elapsed with respect to a certain method of establishing temporal relations α, for any objects a time x elapsed with respect to α. This statement is a logical tautology, but this circumstance is hidden behind a loose verbal formulation. In a different manner (and more explicitly) it can be formulated as follows: for any object a, if time can be measured for a by a method α and if this results in a quantity x, the magnitude of time for a with respect to α is characterized by a quantity x.

12. The Irreversibility of Time

The question of the reversibility and irreversibility of time represents a part of the question of variation of time. We have singled it out here in order to draw the reader's attention to still one more detail.

The question of reversibility and irreversibility of time is frequently confused with that of reversibility and irreversibility of processes. But even if one recognizes these two issues as different aspects of the same problem, the following fact should be taken into account.

If we observe the transformation of A into B, and then the inverse transformation of B into A, this does not correspond to reversal of time: if at t^1 one observes A, and then at t^2 B takes place, the inverse transformation of B into A is possible only at a time t^3 which follows after t^1 and t^2. Besides, the A that has turned into B, and the A into which B has turned do not represent the same object because of the definition of the expression 'the same object'. The transformation $B \Rightarrow A$ results in an object which is identical to A (belongs to the same class as A) but is not the same A.

The irreversibility of time has no physical ground whatsoever. Temporal terminology is developed in such a way (and for such relations between objects) that in view of the very method of formulation of this terminology we must recognize the irreversibility of time in order to avoid conflict with definitions of terms.

Assume that the interval of time t^1 occurs later than t^2 (in other words, all variations taking place during t^1 occur later than all variations taking place during t^2). The meaning of the expression 'time is reversible' (in the simplest case) is the following: there is a possibility of t^1 and t^2 such that the temporal relation for them can be replaced by the reverse (i.e., an arrangement is possible when t^2 occurs later than t^1). In this case the same reference point for time is assumed (which is of no importance here). Let $t^1 > t^2$ mean that t^1 is later than t^2, and $t^2 > t^1$, vice versa, namely, that t^2 is later than t^1. To reverse time means that at some moment we shall have $t^1 > t^2$, and at a different one, $t^2 > t^1$.

Therefore, in order to answer the question, whether or not it is possible to have time reversed, it is necessary to consider temporal intervals (and moments) themselves as individuals, and to establish what is meant by the term 'the time of existence of time'.

We shall not go into all the details of transferring the above definitions over such individuals. Let us just present an obvious result: the time of existence of a given time t is this time t itself. Thus, t^1 does not exist at t^2, and t^2 does not exist at t^1. If t^2 does not exist at t^1, then t^2 will never exist in future. Therefore, at a time t^3, when we would like to obtain $t^2 > t^1$, there will exist neither t^2, nor t^1 (since $t^3 > t^1$). Thus, neither t^1 nor t^2 is reproducible, and the relation between them cannot be changed. The above can be made more convincing if reformulated in such a way that t^1 and t^2 are replaced by clusters of variations at t^1 and t^2 which are individual and therefore not reproducible.

If one makes use of Φ^8, the following assertion concerning the irreversibility of time can be proved:

$$(t^2 > t^1) \wedge (t^3 > t^2) \rightarrow \neg \mathrm{E}t^3 \downarrow (t^1 > t^2).$$

Indeed, if $t^3 > t^2$, then, by virtue of Φ^8, $\vdash \neg \mathrm{E}t^3(t^2)$, and due to the definition of the existence of relation, $\vdash \neg \mathrm{E}t^3(\downarrow (t^1 > t^2))$. The following assertions of the impossibility of time being accelerated or slowed down:

$$(t^2 > t^1) \wedge (t^3 > t^2) \rightarrow \neg \mathrm{E}t^3(\downarrow (t^1 = t^2)),$$
$$(t^2 = t^1) \wedge (t^3 > t^2) \rightarrow \neg \mathrm{E}t^3(\downarrow (t^2 > t^1)),$$
$$(t^2 > t^1) \wedge (t^2 \parallel t^1) \wedge (t^3 > t^2) \rightarrow \neg \mathrm{E}t^3(\downarrow ((t^2 > t^1) \wedge$$
$$\wedge (t^2 \neg \parallel t^1))),$$
$$(t^2 > t^1) \wedge (t^2 \neg \parallel t^1) \wedge (t^3 > t^2) \rightarrow \neg \mathrm{E}t^3(\downarrow ((t^2 > t^1) \wedge$$
$$\wedge (t^2 \parallel t^1))).$$

The term ‘reversibility of time’ is also used when the possibility of changing the direction of the temporal series is implied. Consider the linguistic expressions appearing in a related argument. The direction of a series A is established with respect to another series B of the same kind. The series A does not represent a segement of B, and B is not a segment of A. Comparison of A and B is carried out by a given method. Thus, for logical clarity, we must consider such expressions as ‘the direction of a series A with respect to B’. To speak of a change in series direction, it is necessary to establish its direction at one time and then its direction at a different time with respect to the same α. In order to introduce the difference in directions of two series A^1 and A^2, or of segments A^1 and A^2 of a single series A, one has to compare their directions with respect to the same α. In this case, A^1 should not represent a segment of A^2, and A^2 should not be a segment of A^1. Comparison of directions of a series as a whole and of a segment of the series is meaningless. The temporal series is a series consisting of temporal intervals and moments. The order of the latter is reduced to the order of the changes occurring in them in the following manner: t^2 is superior in order to t^1 with respect to γ if and only if all the changes occurring at t^2 are superior in the temporal order to all the changes occurring at t^1, with respect to β. In order to be able to speak of a change in a temporal series, we have to consider the temporal series with respect to a certain β at a time t^1 and then to consider the same series at a time t^2 after t^1 with respect to β. In this case the temporal series at t^2 will differ from the temporal series at t^1 only by the presence of an interval between t^1 and t^2 and the interval t^2. As a consequence of the adopted conditions and definitions, the temporal series at t^1 represents a segment of a temporal series at t^2, and it would be meaningless to compare their directions. The notion of the difference in directions of two temporal series would make sense if one considers a temporal series in the interval t^1 and a temporal series in a different interval t^2, one of t^1 and t^2 being superior to the other with respect to β. To compare the directions of temporal series at t^1 and t^2, it is necessary to establish correspondence between each of them and the method β. But then, by virtue of conditions specified, the temporal series at t^1 represent a segment of a temporal series with respect to β, and it would be meaningless to speak of its direction. The argument for t^2 is similar. Thus, it makes no sense to compare the direction of these two objects.

13. On the Relation of Generation

Assume that an individual A left the Earth at a time t^1 and returned at a time t^2. According to the very meaning of the word 'returned', the time t^2 follows after t^1 with respect to a certain reference point for time α. If an individual B existed at t^1 or before t^1 and does not exist at t^2, then, by virtue of Φ^3 and Φ^4 this individual will never exist after t^2. Thus, if B represents a father or a mother of the individual A, then at least the following may be said: wherever A went and whatever course he took (and whatever happened to him in his travels), if, after coming back he does not see his parents alive, they will never come into being, our certainty on this point following from the analysis of terms in which our assertions are formulated.

The relation between children and parents represents a special case of the relation 'generates'. By definition, the following is valid for this relation: if a generates b, then for any method of establishing temporal order the individual a appears before b. If α and β are the variables for time, and a and b are the individual variables, the above can be written as follows: if a generates b, then for any method of reading time we have

$$(\exists\alpha)(\exists\beta)(\mathrm{E}\alpha(a) \wedge \neg\mathrm{E}\alpha(b) \wedge \mathrm{E}\beta(b) \wedge (\beta > \alpha)).$$

We are not discussing the meaning of the term 'generates' in detail here. However, the above statement represents a part of its implicit definition. If this definition is not adopted, then either the relation between parents and children cannot be considered as a special case or it is completely meaningless to continue this discussion.

14. The Continuity of Space and Time

The continuity of space and time cannot be reduced to the continuity of a series consisting of empirical objects in space and time. Those are continuous in the following sense: between any spatial (temporal) structures there are also spatial (temporal) structures. Let us make this statement more precise.

First, consider intervals. Let $\{a, b, \alpha\}$ and $\{c, d, \alpha\}$ be the spatial intervals, and $b > \alpha c$. There are two kinds of approach possible. With the first, we accept that under specified conditions $\{b, c, \alpha\}$ also represents a spatial interval. But then the question arises: how to interpret $\{a, d, \alpha\}$? In order to recognize it as a spatial interval, one more axiom is needed (the axiom of the addition of intervals): if $\{a, b, \alpha\}$ and $\{b, c, \alpha\}$ are spatial intervals, then $\{a, c, \alpha\}$ is also a spatial interval. When the second approach is adopted, we assume that for the specified conditions $\{a, d, \alpha\}$ is also a spatial interval. But then a question arises with respect to $\{b, c, \alpha\}$. In order to interpret it as a spatial interval, another axiom is needed (the axiom of the division of intervals); if $a > \alpha\, b$ and $b > \alpha\, c$ and at the same time $\{a, c, \alpha\}$ is a spatial interval, then $\{a, b, \alpha\}$ and $\{b, c, \alpha\}$ are spatial intervals. The argument for time is quite similar. These two approaches are equivalent.

If the axiom of addition of intervals is adopted, then we have to accept the following axiom X^1 by means of which we can define the continuity of space at the level of intervals: if $\{a, b, \alpha\}$ and $\{c, d, \alpha\}$ are spatial intervals and if $b > \alpha c$, then $\{b, c, \alpha\}$ is also a spatial interval. Let us call X^1 the axiom of continuity. Then the definition of the continuity of space will take the form: 'space is continuous' $\equiv$ Df. X^1. The procedure for time is similar.

If we adopt the axiom of division of intervals, the following axiom of continuity X^2 must be adopted: if $\{a, b, \alpha\}$ and $\{c, d, \alpha\}$ are spatial intervals and if $b > \alpha c$, then $\{a, d, \alpha\}$ is also a spatial interval. The continuity of space will be defined as: 'space is continuous' ≡ Df. X^2. The formulation for time is similar.

While in the case of time the above definition appears sufficient (time is understood as 'non-planar' and 'volume-free'), for space we need a more general definition that takes arbitrary spatial structures into account. Note that it is required for time also if the presence of different spaces in the methods of reading time is kept in mind. In any case, we are allowed to see in the definition for time given below only binary structures with the general method of establishing order (an ordinary special case).

The following relations are known for structures: one structure is a substructure of another, and one structure is inside another (if all the elements of the first are inside the other). The second relation may be reduced to the first one in the following manner. If a structure A is inside a structure B with respect to a certain class of methods of establishing order α, then (1) elements of A form a structure A* with respect to α, whereas elements of B form a structure B* with respect to α; (2) elements of A and B form a structure C with respect to α; (3) A* represents a substructure of C and B* represents a substructure of C. Therefore, in what follows it would be sufficient to consider the relation between structure and substructure.

Let us adopt the following axiom of addition of spatial structures Φ^9: if A and B are spatial structures, C is a structure with respect to α, A and B are substructures of C with respect to α, then C is also a spatial structure (i.e., the addition of spaces results in a space, or the sum of spaces is a space). The argument in the case of time is similar (the addition of times results in a time; the sum of times is a time).

Let us adopt the following axiom of continuity of space Φ^{10}: if A and B are spatial structures with respect to α, and C is a structure formed with respect to α of elements of A (with at least one element being taken from A) and of elements of B (also, with at least one element of B being taken), then C represents a spatial structure with respect to α (i.e., there is space between any spaces). The situation for time is similar (there is a time between any times).

A different variant contains the axiom of division of space and time. If the first one is adopted, the second is obtained as its consequence. And vice versa.

Assertions Φ^9 and Φ^{10} represent parts of the implicit definition of space and time as such. They can also be interpreted as a definition of continuity of space and time. It is obvious that the continuity of series in space and time is not the same as the continuity of space and time.

15. Invariance of Space and Time

Space and time are invariant in the sense of the following axiom Φ^{11}: if A is a spatial (temporal) structure with respect to α, and A* is a structure formed of the elements of A with respect to β, then A* is also a spatial (respectively, temporal) structure.

This can be interpreted in such a way that space is always space (and time is always time), regardless of how it is limited, read, measured, etc.

16. Identity and Difference of Location and Time

Consider A and B, spatial structures with respect to α. Statements 'An individual is inside A (inside B) at a time t with respect to α' will be denoted by x and y, respectively. Let us adopt the following definitions:

(1) A and B intersect with respect to α at a time t if and only if $(\exists I)(x \wedge y)$.

(2) A and B coincide with respect to α at a time t if and only if $(\forall I)(x \wedge y)$.

Similar definitions can be formulated for temporal structures, with the only difference that the expression 'a time t' is replaced everywhere with 'a place x'.

It follows from Definitions 1 and 2 that

(3) A and B do not intersect with respect to α at a time t if and only if $(\neg\exists I)(x \wedge y)$.

(4) A and B do not coincide with respect to α at a time t if and only if $(\neg\forall I)(x \wedge y)$.

The assertion 'a physical body cannot occupy different places simultaneously' is accepted by a majority of people. But why is it so? The usual answer is to the effect that this is how nature is built.

But in reality nature has nothing to do with this. And the above assertion does not appear so indisputable if one considers it in detail. Actually, this statement is adopted only inasmuch as one implicitly adopts a certain definition of the terms 'different places' and 'at the same time'.

The expressions 'different places' and 'different times' are ambiguous: they may mean that some spatial (temporal) structures do not intersect, and may mean that they do not coincide. And the expressions 'the same place' and 'the same time' may mean that spatial (temporal) structures intersect and may mean that these structures coincide.

Besides, the identity of and the difference between locations and times is understood as the identity of and the difference between objects with respect to spatial and temporal order, i.e., as the identity of and the difference between the points of space and the moments of time in the spatial and temporal order of objects. Thus, for instance, the expression '*a* at the same time as *b*' may mean $a = \alpha b$, where α is a method of reading time, and also that the temporal intervals in which *a* and *b* exist coincide.

If one adopts a definition, according to which A and B represent different locations if and only if they do not intersect, the following consequence is obtained from assertion (3): an empirical individual cannot simultaneously occupy different locations. The word 'simultaneously' is understood here in the sense of the identity of moments or coincidence of temporal intervals. If, on the other hand, different locations are defined in the 'weaker' sense, just as non-coinciding places, or if different times are defined as non-coinciding temporal intervals, there is a possibility for an individual to occupy different locations at the same time (if it is placed within their common part).

17. The Predication of Variations

The variation $\downarrow x \Rightarrow \downarrow y$ can be considered as a special object $\downarrow(\downarrow x \Rightarrow \downarrow y)$ and is assigned predicates. Sometimes the latter should be defined separately, namely, when we take into account the peculiarities of this logical type of the empirical object.

Variation is considered as discrete if the transition state is not taken into account. If it is assumed that there is no transition state at all, abstract variation is allowed. Variation is considered as non-discrete if the transition state is taken into account, and the latter, in turn, is regarded as a set of variations, etc. until one reaches a continuous series of intermediate changes. Thus, the predicates 'discrete' and 'non-discrete', when applied to variation, acquire an additional nuance (in comparison with their application to series), namely, specification of the method of analysis.

The variation $\downarrow(\downarrow x \Rightarrow \downarrow y)$ exists (is realized) in the interval of time t if and only if the following are valid:

(1) at $t^1 \downarrow x$ exists (and $\downarrow y$ does not exist);
(2) later at a time $t^2 \downarrow (\sim x \wedge \sim y)$ exists;
(3) later at a time $t^3 \downarrow y$ exists (and $\downarrow x$ does not exist);
(4) t^1, t^2, t^3 form a continuous series;
(5) t represents a temporal interval which includes the interval between t^1 and t^3, as well as the parts of t^1 and t^3 that are adjacent to it and constitute a necessary and sufficient condition for the existence of $\downarrow x$ and, respectively, of $\downarrow y$; t is the proper time of $\downarrow(\downarrow x \Rightarrow \downarrow y)$.

The variation $\downarrow(\downarrow x \Rightarrow \downarrow y)$ exists only at the proper time t. As soon as it has occurred, it stops existing as a separate individual. It does not exist as an individual at any time before t^1 nor at any time after t^3.

The predicate $\Rightarrow$ is a two-place one. But there is also a one-place predicate 'varies', 'has varied', and so on (it will be denoted by a symbol $\Downarrow$), which can be defined in the following way: $\Downarrow(a)$ if and only if a certain statement is possible, containing a as a subject, that is true at one time and not-true at a different time; this statement is neither E(a) nor $\urcorner$E(a), i.e., the existence of a is implied.

Thus, the predicate $\Downarrow$ is just a modification of $\Rightarrow$ for special cases.

Can one apply the predicates of variation to the variation itself? This question is far from idle, since various kinds of speculation are associated with lack of clarity in such cases of the usage of terminology.

What can result from application of predicates $\Rightarrow$ and $\Downarrow$ to the variation $\downarrow(\downarrow x \Rightarrow \downarrow y)$? We have the following opportunities: (1) $\downarrow x$ turns not into $\downarrow y$ but into $\downarrow z$; (2) $\downarrow x$ turns into $\downarrow y$, but in a different way. In the first case another variation, namely, $\downarrow(\downarrow x \Rightarrow \downarrow z)$ occurs. In the second case a question arises: in a different way as compared to what? If $\downarrow(\downarrow x \Rightarrow \downarrow y)$ has occurred, it does not exist already, and will never exist. Therefore, one can compare only cases of variation belonging to the same class.

Assume that $\downarrow x^1 \in K \downarrow x$, $\downarrow x^2 \in K \downarrow x$, $\downarrow y^1 \in K \downarrow y$ and $\downarrow y^2 \in K \downarrow y$. Assume that variations $\downarrow(\downarrow x^1 \Rightarrow \downarrow y^1)$ and $\downarrow(\downarrow x^2 \Rightarrow \downarrow y^2)$ can or do occur. These variations represent elements of the same class of variations $K\downarrow(\downarrow x \Rightarrow \downarrow y)$. They may differ in the most diversified attributes in location, in duration, etc. But this is not a variation of both $\downarrow(\downarrow x^1 \Rightarrow \downarrow y^1)$ and $\downarrow(\downarrow x^2 \Rightarrow \downarrow y^2)$.

Thus, it is simply impossible to apply the predicates of variation to variations themselves, since the corresponding statements are meaningless.

One may formulate assertions that constitute a part of the implicit definition of the predicate of variation, in particular, the following

(1) $$\vdash (\downarrow x \Rightarrow \downarrow y) \rightarrow \sim (\downarrow x \Rightarrow \downarrow \sim y),$$

(2) $\vdash (\downarrow x \urcorner \Rightarrow \downarrow y) \leftrightarrow \sim (\downarrow x \Rightarrow \downarrow y)$,

(3) $\vdash \sim (\downarrow x \Rightarrow \downarrow y) \rightarrow x \vee ((\downarrow x \Rightarrow \downarrow z) \wedge \sim (y \wedge z))$,

(4) $\vdash \sim y \rightarrow \sim (\downarrow x \Rightarrow \downarrow y)$,

(5) $\vdash x \rightarrow \sim (\downarrow x \Rightarrow \downarrow y)$,

(6) $\vdash (\downarrow x \Rightarrow \downarrow y) \rightarrow (\downarrow x \Rightarrow \downarrow \sim x) \wedge (\downarrow \sim y \Rightarrow \downarrow y)$.

The last of these reads as 'if $\downarrow x$ turns into $\downarrow y$, then $\downarrow x$ disappears and $\downarrow y$ emerges at the same time'. A special case: 'If an object *a* turns into *b*, then at the same time *a* disappears and *b* emerges'.

18. Translation, Change of Place (Movement)

Translation of bodies in space (motion) represents a special case of variaton.

The predicate $\Rightarrow$ in the case of translation has the following meaning. Assume that x is a statement 'A body *a* occupies a location α', y is a statement 'A body *a* occupies a location β', α and β are different locations, the state $\downarrow x$ precedes $\downarrow y$ in time. The formula $\downarrow x \Rightarrow \downarrow y$ reads briefly as 'A body *a* has moved (is moving) from location α to location β'.

The expression $\Downarrow(a)$ ('A body *a* moves') is ambiguous. It may mean that there is a transition state $\downarrow(\sim x \wedge \sim y)$. But it can be explicated as follows: at a given time t a spatial domain α can be specified such that both assertions 'A body *a* is in α' and 'A body *a* is not in α' are wrong.

The expression '*a* has moved from α to β' means, by definition, that at a certain time t^1 an object *a* has occupied a location α, and at a time t^2 – a location β, and that $t^2 > t^1$. The expression '*a* has not moved from α (is still in α)' means that *a* has been in α both at t^1 and at t^2, and at any time between t^1 and t^2 the object *a* was in α.

But then the question arises: is it possible that one registering device or one person A has recorded that *a* occupies α at a time t, while another registering device or another person B has recorded that *a* occupies another place β at the same time t? Let us recall the definition of different locations. According to this definition, if α and β represent non-intersecting locations. then from the point of view of the same method of establishing order *a* cannot occupy different locations simultaneously. Therefore, under the condition that a difference between A and B means a difference in the methods of establishing the order, we cannot regard α and β as different locations. And if the difference between A and B represents a difference in the frame of the same method of establishing order, the above situation is impossible.

If we adopt the definition of different locations as non-intersecting (having no 'points' in common), the impossibility of instantaneous translations follows logically from it. A translation of a body a from location α into another location β is considered instantaneous if and only if at a time t, a occupies α and at the same time occupies β. But if, according to the definition of different locations, no object a can occupy α and β at the same time, instantaneous translation is impossible. Thus, the assumption of the existence of instantaneous translations is a purely physical assumption. It is legitimate only if interpreted as an intention to ignore time, spent on translation in one or another case, or as an assumption of the existence of abstract objects that can move outside of time.

19. The Paradox of Motion

Motion, i.e., translation of bodies in space, represents only a special case of variation of objects. Everything that has been said above concerning variations in general can be applied to motion. Below we shall consider the well-known paradox of motion: 'A moving body occupies and at the same time does not occupy a given location of space'.

The question 'Can a physical body occupy and at the same time not occupy a given location?' is answered in most cases in the negative. And in most cases the motivation for the negative answer deserves criticism.

A physical body cannot occupy and at the same time not occupy a given location because this is how the world is made – this is a frequent answer to the above question. Indeed, our experience does not contain cases that contradict such an answer. And it never will. But the cause of this is different in principle from the causes of the fact that there are no horses with ten horns or hares with hooves. The reason for our situation is that we make use of the signs 'and' and 'not', while the statement 'A physical body occupies at a given time a given location' represents a special case of statements $\vdash \sim (x \wedge \sim x)$. And there is no other meaning here.

The difficulty in eliminating any paradoxes lies first of all with the necessity of providing a rigorous description of the method of arriving at them. This is also the situation in our case. Where does the assertion 'A moving body does and at the same time does not occupy a given location' come from? Does it result from empirical observation? No. What is impossible logically is impossible in reality, and the impossible (and the non-existent) cannot be observed. Therefore, it is either accepted as an axiom or follows as a consequence from other statements. Since it is logically contradictory (not-true by virtue of the properties of conjunction 'and' and negation 'not'), it cannot be accepted as an axiom. Therefore, it must be regarded as a consequence of some other assumptions. Which exactly?

There are empirically observed cases when one cannot say about a moving object that it occupies a certain location, and one cannot say that it does not occupy it, i.e., when a transition state takes place. If P(*a*) is a statement '*a* occupies α', with respect to the transition state, the assertion

(1) $\sim P(a) \wedge \sim \urcorner P(a).$

is true. However, if the two negations are not distinguished, i.e., one does not discriminate between $\sim$ and $\urcorner$, and the latter is perceived as $\sim$, then the transition state is erroneously described by an expression

(2) $\sim P(a) \wedge \sim\sim P(a).$

By virtue of the validity of the rules of logic

(3) $\sim\sim P(a) \dashv\vdash P(a),$

it follows from (2) and (3) that

(4) $\sim P(a) \wedge P(a).$

And if we accept (2), we have to accept (4) also. Hence, according to rules of logic,

(5) $(\exists x)(\sim P(x) \wedge P(x)),$
(6) $\sim(\forall x)(P(x) \vee \sim P(x)),$

where x is an individual variable, i.e., the law of excluded third $\vdash A \vee \sim A$ is not valid for all individuals.

The paradox is caused by an error, the essence of which is the following: from the very beginning we assume the non-classical case, with the assertions

$$\vdash P(a) \vee \urcorner P(a),$$
$$\sim P(a) \vdash P(a),$$

being wrong, since three possibilities are implied:

$$P(a), \urcorner P(a), \sim P(a) \wedge \sim \urcorner P(a),$$

and at the same time we apply the rules for the classical case for which the negations $\neg$ and $\sim$ are not distinguished, and therefore the above assertions are true, since only two possibilities

$$P(a), \neg P(a)$$

are implied.

20. Process

A process is a series of variations, ordered in time and possessing the following properties: assume that $a \Rightarrow b$ and $b \Rightarrow c$ is an arbitrary pair of adjacent elements of this series; if the state a exists at t^a, the state b at t^b, and the state c at t^c, then $t^c > t^b$ and $t^b > t^a$. The states a, b, c represent states of the same individual A.

The problem of the existence of a process is solved in accord with the definition of the term denoting it. For instance, a process B may be defined in the following way: it is a sequence of variations of an object A such that $(a \Rightarrow b) \wedge (b \Rightarrow c)$. Obviously, B exists if and only if both $a \Rightarrow b$ and $b \Rightarrow c$ exist in the corresponding temporal order. But the process B can be called 'a process of the individual A'. In order for such process to exist, it is sufficient that at least one variation of the individual A exists. To put it differently, the question of the existence of a process is answered by taking into account the type of term denoting it.

There are paradoxes associated with processes. A process having no beginning does not start and, therefore, does not exist (does not occur). However, it is believed that the process of variations in the world had no beginning (is infinite in the past), but at the same time it does exist. If a process consists of an infinite number of variations, it does not end. It is believed, however, that any process that has ended may be infinitely decomposed into constituent variations.

To resolve these paradoxes, one has first of all to have in mind that the expression 'a process has no beginning (does not start)' is ambiguous. A process can have no beginning if considered as a series of variations without the starting element. And in a different situation, a process can have no beginning in the sense that the individual exists all the time (i.e. a given time, or time in general) in a certain state, and the latter does not turn into any other state. For instance, the process consisting of variations $a \Rightarrow b$ and $b \Rightarrow c$ has no beginning from the point of view of the second definition, if the state a is preserved at times t^b and t^c.

Obviously, if a process has no starting element (i.e., for any of its elements

α there is another element β preceding it), it does not follow that it does not exist. On the other hand, if the process has not started (i.e., a certain state of an individual has not been replaced by other states) then, according to the definition of the existence of a process, such a process does not exist. The first paradox results from a confusion of notions, i.e., from an ordinary logical error.

As for the second paradox of process, it would be instructive to consider the following simplest case. Assume that a variation $a \Rightarrow b$ has taken place. Its analysis may reveal that this variation represents a process consisting of variations

$$a \Rightarrow c^1, c^1 \Rightarrow c^2, \ldots, c^n \Rightarrow b.$$

And it appears that a similar analysis may be applied to each variation of this process, so that every variation can be presented as a process consisting of an infinite number of variations. But each variation requires time. And if the latter is greater than zero and finite, an infinite number of variations may be required, and there may be no end to this process. Therefore, the variation $a \Rightarrow b$ is impossible. But the same may be said about each variation of a given process.

The cause of this paradox is in a confusion of abstract and empirical processes. In the case of abstract processes, we admit the existence of variations that do not require time or can be arbitrarily fast. Then the results of reasoning for abstract processes are transferred over to empirical processes, which is not always legitimate, and in this case wrong.

21. Minimal Dimensions

With respect to abstract objects, assumptions may be made according to which the dimensions of objects may be equal to zero, while for each interval between objects which is greater than zero there is another interval that is shorter than the first one and greater than zero. There is no danger in such assumptions if we keep in mind that they pertain only to abstract objects. If x and y are variables for intervals, the above assumptions can be written as

(1) $(\exists x)(x = 0)$,

(2) $(\forall x)(\exists y)((x > 0) \wedge (y > 0) \wedge (x > y))$.

The situation with empirical individuals is different. The term 'the interval between a and b with respect to α' is meaningful only if $a > \alpha b$

or $b > \alpha a$, in other words, if a and b represent elements of an ordered series. If a and b are empirical variations, and the interval between them is supposed to be temporal, the positions of a and b in the temporal series do not change. On the other hand, if a and b represent empirical bodies, and the interval between them is supposed to be spatial, one has to assume the existence of a fixed time during which the interval does not vary. When we speak of the variation of an interval over a certain time, we imply a temporal series of intervals with each element fixed. Thus, under these conditions only the following cases are left. Case 1: a is in contact with b, i.e., no other individual may be placed between them. They are moved up against each other. The distance between them cannot be further decreased. In this case it is the minimal. Case 2: a is not in contact with b, i.e., another individual can be placed between them. The distance between these objects may be decreased; it is greater than the minimal one. Case 3: a is or is not in contact with b, and we do not know exactly what is the situation. Since it is not our cognitive possibilities which we consider here but rather the relation between a and b regardless of our presence, we shall be justified to treat a and b in this case as either being or not being in contact, i.e., to exclude uncertainties. To put it differently, as far as terms for intervals between empirical individuals are concerned, one can adopt, as part of their implicit definition, the following assertion:

Φ^{12}. $\quad \vdash (a \parallel \alpha b) \vee (a\urcorner \parallel \alpha b)$.

Instead of the expression 'The dimensions (length, duration) of the minimal interval belonging to the class of intervals between empirical individuals' we shall use the expression 'min'. A similar abbreviation will be employed below for the class of dimensions of empirical individuals.

Taking the above into account, let us adopt the following definitions:

Φ^{13}. $\quad (l\{a, b, \alpha\} = \text{min}) \dashv\vdash (a \parallel \alpha b)$,
Φ^{14}. $\quad (l\{a, b, \alpha\} > \text{min}) \dashv\vdash (a\urcorner \parallel \alpha b)$.

The first of these reads as 'The interval between a and b is called minimal if and only if a is in contact with b'. The meaning of the second one is 'the interval between a and b is called exceeding the minimal one if and only if a is not in contact with b'.

Consider now the following trivial argument. The substitution of Φ^{13}, Φ^{14} for Φ^{12} results in

(1) $\quad \vdash (l\{a, b, \alpha\} = \text{min}) \vee (l\{a, b, \alpha\} > \text{min})$.

According to the rules of the logic of relations, we obtain:

(2) $(l\{a,b,\alpha\} = \min) \dashv \sim (l\{a,b,\alpha\} < \min)$,

(3) $(l\{a,b,\alpha\} > \min) \dashv \sim (l\{a,b,\alpha\} < \min)$.

It follows from (1)–(3) that

(4) $\vdash \sim (l\{a,b,\alpha\} < \min)$.

Since the above is valid for any interval, by the rules for introduction of a variable z for terms of intervals between empirical individuals we obtain from (1) and (4):

(T[1]) $\vdash (\forall z)(lz \geqslant \min)$,

$\vdash (\forall z) \sim (lz < \min)$.

Applying the rules for quantifiers to these expressions results in

$$\vdash (\neg\exists z)(lz < \min).$$

As follows from the rules for relation between the quantifier $\exists$ and the predicate E.

$$\vdash \neg E(z\downarrow (lz < \min)).$$

Assume, further, that x and y represent variables for individual terms. It is provable in the logic of relations that

$$\vdash (lx = ly) \vee (lx > ly) \vee (lx < ly) \vee (lx ? > ly).$$

Let z be a term of interval such that $lz = \min$. Such a term may be constructed. For instance, it can be $\{a,b,\alpha\}\downarrow (l\{a,b,\alpha\} = \min)$, in other words, the term 'An interval between a and b with respect to α such that its dimensions are minimal'. By making use of the rule $\vdash P(a\downarrow P)$ from the theory of terms ('An object, having an attribute, has this attribute'), we obtain the result: an interval, having minimal dimensions, has minimal dimensions. Thus, if the term z is constructed in such a way that $lz = \min$, then the following will be valid:

(1) $\vdash lz = \min$.

After substituting z in place of x in the formula from the logic of relations we obtain

(2) $\vdash (lz = ly) \vee (lz > ly) \vee (lz < ly) \vee (lz ? > ly)$.

Further, we have:

(3) $\vdash (lz > ly) \wedge (lz = \min) \rightarrow (ly < \min)$.
(4) $\vdash \sim (ly < \min) \rightarrow \sim (lz > ly) \wedge (lz = \min))$.

Since

(5) $\vdash \sim (ly < \min)$,

it follows from (4) and (5) that

(6) $\vdash \sim (lz > ly) \wedge (lz = \min))$.

On the basis of (6) and (1), we have:

(7) $\vdash \sim (lz > ly)$.

From (2) and (7) we obtain:

(8) $\vdash (lz = ly) \vee (lz < ly) \vee (lz? > ly)$.

As follows from the rules of the logic of relations,

(9) $(lz = ly) \vdash \sim (lz > ly)$.
(10) $(lz < ly) \vdash \sim (lz > ly)$.
(11) $(lz? > ly) \vdash \sim (lz > ly)$.
(12) $\vdash \sim (lz > ly)$.

Applying the generalization rule, we have:

(13) $\vdash (\forall y) \sim (lz > ly)$.

Finally, from the rule for introduction of a variable it follows that

(T^2) $\vdash (\exists x)(\forall y) \sim (lx > ly)$,

in other words, a statement concerning the interval the dimensions of which do not exceed the dimensions of any other interval.

Let us recall that the definition of dimensions of an empirical individual reduces to the definition of dimensions of an interval, i.e.,

$$\vdash (l\{c, \alpha\} = l\{a\downarrow x, b\downarrow x, \alpha\}),$$

where x is $(a \| \alpha c) \wedge (c \| \alpha b)$.

Since x is true, we have

$$\vdash \sim (a \| \alpha b),$$
$$\vdash (l\{a\downarrow x, b\downarrow x, \alpha\} \geqslant \min),$$
$$\vdash (l\{a, b, \alpha\} \geqslant \min),$$

and, therefore,

$$\vdash l\{c,\alpha\} \geqslant \min.$$

Since c represents an arbitrary empirical individual, we have

$$(\mathrm{T}^3) \qquad \vdash (\forall x)(l\{x,\alpha\} \geqslant \min), \quad \vdash (\forall x) \sim (l\{x,\alpha\} < \min),$$

where x is the individual variable (for empirical individuals, certainly). Application of the rules for quantifiers and predicates $=$, $>$ and $<$ results in

$$\vdash (\neg\exists x) \sim (l\{x,\alpha\} \geqslant \min), \quad \vdash (\neg\exists x)(l\{x,\alpha\} < \min).$$

Thus, we have obtained the result that the dimensions of any empirical individual with respect to a given method of establishing order α cannot go below a certain minimum. Since α represents an arbitrary method of establishing order, we obtain

$$\vdash (\forall\alpha)(\forall x) \sim (l\{x,\alpha\} < \min).$$

Now, let us consider an interval $\{a,b,\alpha\}$ which is greater than the minimal, and move a closer to b. This move is equivalent to a choice of an interval $\{c,d,\alpha\}$ which is shorter than $\{a,b,\alpha\}$. When a is moved closer to b (or when the intervals are compared), we have the following situation. Either this move has no effect on the dimensions of the interval (the same individual or the same number of certain individuals as before is placed between a and b) or the dimensions of the interval will change its magnitude in a discontinuous manner, by the length of one individual: there will be no place for it between a (or b) and the individuals located between a and b, through which the dimensions of a, b, α are defined. And because of the fact that the dimensions of individuals cannot be smaller than a certain minimum, moving a closer to b will represent a series consisting of a finite number of discontinuous stages of the decrease in the interval dimensions until it reaches its minimum (i.e. then it is impossible to insert any individual between a and b). Thus, if the dimensions of the interval x exceed those of the interval y only by a minimal length of a certain individual, and the dimensions of y exceed the dimensions of the minimal interval also by the length of one individual, for a given y there will be no interval z such that $(ly > \min) \wedge (lz > \min) \wedge (ly > lz)$. If $ly > lz$, then under these conditions $lz = \min$. Thus, we have

$$(\mathrm{T}^4) \qquad \vdash (\exists x)(\forall y)((l\{x,\alpha\} \leqslant l\{y,\alpha\}) \wedge (l\{x,\alpha\} > 0)).$$

The above pertains not only to lengths but also to durations. It follows from this discussion that the spatial and temporal relations between empirical individuals, as well as their dimensions, experience discontinuous variations: they vary by quanta (we shall return to this question later).

The following circumstance has to be paid special attention. Since in the general case, from such statements as $(\exists a)w$ we cannot obtain statements of the form $\mathrm{E}(a\downarrow w)$, the assertions ($T^3$) and ($T^4$) cannot be interpreted verbatim as

$$\vdash \mathrm{E}(x\downarrow(\forall y)\sim(lx > ly)),$$
$$\vdash \mathrm{E}(x\downarrow(\forall y)((l\{x,\alpha\} \leqslant l\{y,\alpha\}) \wedge (l\{x,\alpha\} > 0))).$$

We can interpret them only as a possibility of a choice, satisfying the condition which is formulated after the arrow.

Let us adopt the definitions:

Φ^{14}. $(l\{a,\alpha\} = \min) \dashv\vdash (\forall z)((lz < l\{b\downarrow k, c\downarrow k,\alpha\}) \rightarrow (lz = \min)),$
$(l\{a,\alpha\} > \min) \dashv\vdash (\exists z)((lz < l\{b\downarrow k, c\downarrow k,\alpha\}) \rightarrow (lz > \min)),$

where a is a variable for the terms of empirical individuals, z is a variable for the terms of empirical intervals, k is $(b\|\alpha a) \wedge (a\|\alpha c)$.

A theorem, similar to (T^4), can be formulated for empirical individuals.

22. On Infinite Dimensions

Are there empirical individuals and intervals characterized by infinite dimensions? (are they possible or not?) Since the problem of dimensions of individuals may be reduced to that of the dimensions of intervals, it would be sufficient to consider the case of intervals.

The answer to the above question should be represented by such statements as $\mathrm{E}(a\downarrow x)$ and $\mathrm{M}(a\downarrow x)$, where a is a term for interval, and x is a given additional condition (in our case, 'The interval has infinite dimensions') or its negation. However, in order to accept or reject such statements, we must at least know the term a itself. Then, it is necessary to have the definitions of the predicates E and M for terms of intervals. Since the values of these predicates for general terms reduce to that for individual terms, and all derived terms of intervals reduce to the elementary terms of form $\{c,d,\alpha\}$, in order to solve the above problem we have, in the final analysis, to consider such statements as $\mathrm{E}(\{c,d,\alpha\})$, $\mathrm{M}(\{c,d,\alpha\})$, $\mathrm{E}(\{c,d,\alpha\}\downarrow x)$, $\mathrm{M}(\{c,d,\alpha\}\downarrow x)$ together with their negations, where c and d are terms of empirical individuals.

But such an approach to the problem implies that we should distinguish rigorously the cases wherein instead of the terms for empirical individuals, explicit or implicit variables are used (similar to the expressions 'one individual', 'another individual', etc.) from those cases wherein the individual terms are actually applied to empirical bodies and variations. For instance, the expression 'Consider the empirical individuals a and b that are infinitely remote from each other with respect to α' that is equivalent to the assumption of the existence of an interval $\{a, b, \alpha\}$ characterized by infinite dimensions does not mean at all that an individual term for the interval $\{a, b, \alpha\}$ has been constructed, since a and b in this case are not individual terms. Here we confront an illusion similar to the case which arises when we say: 'Consider a true statement $x \wedge \sim x$'. This sentence does not make $x \wedge \sim x$ a concrete statement, nor does it make true any of the statements such as $x \wedge \sim x$. In this situation, in order to give an example of an interval, we have to consider intervals like those between Moscow and Leningrad, between the Sun and the Earth, between the beginning of the First World War and the beginning of the Second, etc.

There is no logical relation whatsoever between the assumption of the infinity of the World in space and time, on one hand, and assertions of the existence (possibility) or non-existence (impossibility) of intervals having infinite dimensions (in space or in time), on the other. The point is that the above assumption means that we suppose that there are no starting or finite elements (or both) in temporal and spatial series. And, as follows obviously from the very structure of terms $\{c, d, \alpha\}$, in the case of intervals their boundaries are specified. And if these intervals are filled up to obtain series, it is exactly the series having a beginning and an end that result.

Predicates E and M have not yet been defined for individual terms such as $\{c, d, \alpha\}$. Since we are discussing spatial and temporal intervals, the definitions should account for certain differences between them. Therefore we adopt the following definitions. For temporal intervals:

$$\mathrm{E}(\{a, b, \alpha\}) \dashv\vdash (\exists t^1)(\exists t^2)(\mathrm{E}t^1(a) \wedge \mathrm{E}t^2(b) \wedge (a > \alpha b)),$$
$$\mathrm{M}(\{a, b, \alpha\}) \dashv\vdash (\exists t^1)(\exists t^2)(\mathrm{M}t^1(a) \wedge \mathrm{M}t^2(b) \wedge (a > \alpha b)),$$

where a and b represent variables for terms of empirical variations, t^1 and t^2 are variables for terms of time (for time 'markings'). For spatial intervals:

$$\mathrm{E}t(\{a, b, \alpha\}) \dashv\vdash (\exists s^1)(\exists s^2)(\mathrm{E}s^1 t(a) \wedge \mathrm{E}s^2 t(b) \wedge (a > \alpha b)),$$
$$\mathrm{M}t(\{a, b, \alpha\}) \dashv\vdash (\exists s^1)(\exists s^2)(\mathrm{M}s^1 t(a) \wedge \mathrm{M}s^2 t(b) \wedge (a > \alpha b),$$

where a and b represent variables for terms of empirical bodies, s^1 and s^2 are variables for spatial domains (for positions of bodies in space), t is a term (or a variable) for time. In the definitions for the temporal interval, t has been omitted, since the time of existence of a temporal interval is this interval itself. Thus, in order for an individual empirical interval to exist, it is necessary that the empirical individuals forming its boundaries do exist. One can say, even without further formal analysis, that statements concerning the existence (possibility) of an individual which is at an infinite distance from a certain chosen individual in space and time can be neither proved nor disproved by experiment. What happens to these statements depends on the system of definitions, i.e., it is related at least partially to the state of the language.

Finally, we need definitions for the expressions 'has finite dimensions' and 'has infinite dimensions' for the terms of intervals. And in this case there is a possibility of diverse variants that are far from being always equivalent.

Let us adopt the following definitions:

D1. An empirical continuous series consisting of individuals with minimal (finite) dimensions is characterized by finite dimensions if and only if it has both starting and final elements, and by infinite dimensions if and only if it has no starting or final element (or both).

D2. An empirical series A consisting of elements characterized by the minimal (finite) dimensions will be called intermediate for $\{a, b, \alpha\}$ if and only if it is continuous, all its elements are between a and b, its starting element is in contact with b, and its final element is in contact with a (with respect to α).

D3. An interval $\{a, b, \alpha\}$ has finite (infinite) dimensions if and only if there is an intermediate series A for it with respect to α which has finite (infinite) dimensions.

It follows from the above definitions, that any series A, intermediate $\{a, b, \alpha\}$, is characterized by finite dimensions, since it consists of individuals that are minimal (finite) in dimensions, is continuous, and has both a beginning and an end; therefore, in this case $\{a, b, \alpha\}$ has finite dimensions; on the other hand, if there is no intermediate series, a statement to the effect that $\{a, b, \alpha\}$ is characterized by infinite dimensions will be wrong.

A weaker definition of the infinite dimensions of an interval $\{a, b, \alpha\}$ may be adopted, namely, the following one: an interval $\{a, b, \alpha\}$ has infinite dimensions if and only if a continuous series A of infinite cardinality can be placed between a and b. In this case the following argument is legitimate. In order to form a judgment concerning the existence of a temporal interval $\{a, b, \alpha\}$ characterized by infinite dimensions, we must first choose an empirical variation b that has occurred at a time t^1. Then another empirical variation a should be chosen such that $a > \alpha b$, and at the same time there is an infinite number of variations occurring between b and a (condition W). The duration of each of the above variations is greater than zero, and this duration is constant, so that the past cannot be changed. Thus, the time during which these variations occur represents all the time after b with respect to α. Let us divide this time after t^1 into finite successive intervals $t_1, t_2, \ldots$ forming a continuous series. If the variation a is chosen at t_1, it obviously does not satisfy the condition W as the number of variations within each finite t_i is finite. Assume that an arbitrary variation a, chosen at t_n, does not satisfy the condition W. But then this condition will not be satisfied by an arbitrary variation at t_{n+1} either, since the sum of the numbers of variations at $t_1, \ldots, t_{n+1}$ is finite. It follows from this that a variation a, chosen within arbitrary t_i, does not satisfy the condition W, in other words,

$$\vdash (\forall t^2) \neg \mathrm{E}t^2(a\!\downarrow \mathrm{W}),$$

where t^2 is a variable for time intervals after t^1. Hence

$$\vdash \sim (\exists t^2)\mathrm{E}t^2(a\!\downarrow \mathrm{W}),$$
$$\vdash \sim ((\exists t^1)(\exists t^2)(\mathrm{E}t^1(b) \wedge \mathrm{E}t^2(a\!\downarrow \mathrm{W}) \wedge (a > \alpha b)),$$

and, finally,

$$\vdash \sim \mathrm{E}(\{a\!\downarrow \mathrm{W}, b, \alpha\}).$$

We would like to point out the fact that this result is equivalent to the assertion 'It is not true that a temporal interval of infinite dimensions exists', but not to the assertion 'A temporal interval of infinite dimensions does not exist'. Since

$$\sim \mathrm{P}(d) \dashv\vdash \neg \mathrm{P}(d) \vee ?\mathrm{P}(d),$$

we obtain

$$\vdash \neg \mathrm{E}(\{a\!\downarrow \mathrm{W}, b, \alpha\}) \vee ?\mathrm{E}(\{a\!\downarrow \mathrm{W}, b, \alpha\}),$$

in other words, either the temporal interval of infinite dimensions does not exist or it is impossible to establish whether it does exist or not.

Next consider spatial intervals. In order to obtain a judgment concerning the existence of an infinitely long spatial interval $\{a, b, \alpha\}$, we must choose an empirical body b existing at a time t in a certain spatial domain s^1, and another empirical body a in another spatial domain, such that $a > \alpha b$. However, in this case a condition W is attached: there should be a continuous infinite series of empirical bodies between a and b, the length of each being greater than zero, and this length is constant over t (if the length of a body varies, the requirement of the continuity of the intermediate series makes it necessary for the empty spaces to fill, and for the extra lengths to be pushed out, so that variation of bodies is of no consequence). We shall divide the whole space beyond b into finite intervals $s_1, s_2, \ldots, s_n$ forming a continuous series. By applying a method of reasoning, similar to the above in the case of time, one can make certain that there is no object a satisfying the condition W in any of the intervals $s_1, s_2, \ldots, s_n$. Thus, we have

$$\vdash (\forall s^2) \urcorner \mathrm{E}s^2 t(a \downarrow \mathrm{W}),$$

where s^2 is a variable for s_i. Hence

$$\vdash \sim (\exists s^2)\mathrm{E}s^2 t(a \downarrow \mathrm{W}),$$
$$\vdash \sim ((\exists s^1)(\exists s^2)((\mathrm{E}s^1 t(b) \wedge \mathrm{E}s^2(a \downarrow \mathrm{W}) \wedge (a > \alpha b))).$$

Thus, under the condition W the infinitely long interval is stylistically unimaginable (simply cannot be cited, or given as an example), since after designating the first boundary of the interval, under the condition W we cannot designate its second boundary in the sense that no empirical individual (body or variation) can be chosen at an infinite distance from a given chosen individual.

As was already mentioned, the illusion to the effect that an infinitely long interval is imaginable, appears to be a consequence of the possibility of constructing linguistic expressions in an arbitrary way, and to ascribe arbitrary properties to them. In this case one can also say: 'Consider an individual a at an infinite distance from a given individual b'. We have already made clear that this does not make the expression 'an individual a' an individual term, denoting an empirically existing individual. We shall cite a curious example of the consequences that result from this expression being confused with an individual term. Let us choose an individual b. Consider a series A such that (1) it has a starting element which is in contact with b with respect to α; (2) it has no final element, in other words,

(1) $(\forall c)(\exists d)((c \in \mathrm{A}) \wedge (d \in \mathrm{A}) \rightarrow (d > \alpha c))$;

(3) every individual that is superior in order to b with respect to α represents an element of A, i.e.,

(2) $(\forall c)((c > \alpha b) \rightarrow (c \in \mathrm{A}))$.

Further, we declare the following: consider an individual a such that $a > \alpha b$ and at the same time all the elements of A are between a and b, in other words,

(3) $(\forall c)((c \in \mathrm{A}) \rightarrow (a > \alpha c) \wedge (c > \alpha b))$.

Thus, it seems that we have conceived of an infinitely long interval $\{a, b, \alpha\}$. However, as follows from (1)–(3),

$$\vdash (a > \alpha b) \rightarrow (a \in \mathrm{A}),$$
$$\vdash (a > \alpha b),$$
$$\vdash (a \in \mathrm{A}),$$
$$(\forall c)((c \in \mathrm{A}) \rightarrow (a > \alpha c)) \wedge (a \rightharpoondown c) \vdash ((a \in \mathrm{A}) \rightarrow (a > \alpha a)).$$

(4) $\vdash (a > \alpha a)$,

which is wrong. In order to avoid this conclusion, we must either acknowledge the fact that postulating a is not equivalent to the choice of an individual or assume that there may not be a series without an end (and a beginning) between individuals a and b.

23. Rate

The rate of a process A in a temporal interval $\{t^1, t^2\}$ of finite dimensions is the number m of consecutive variations of this process, which are pairwise in contact, in this temporal interval.

Obviously, $m \geqslant 1$. If $m = 0$, then variations do not occur, and the process does not exist. The expression 'The rate of a process A in $\{t^1, t^2\}$ equals zero' is just a replacement of the expression 'The process A does not occur (does not exist)'. The above represents a special case of the application of the rules for quantities.

The magnitude of $\{t^1, t^2\}$ is greater than or equal to the minimal. It is greater than zero, otherwise there is no process. And it is finite, otherwise the counting of the number of variations in the process goes on without end. If the process A terminates itself within $\{t^1, t^2\}$, it is self-evident. If A is infinite, and at the same time $\{t^1, t^2\}$ is infinite (in other words, t^2

does not arrive), the number m cannot be obtained, and the concept of rate becomes meaningless. Therefore, the number of variations within $\{t^1, t^2\}$, however, great, is always finite. In other words, as $\{t^1, t^2\}$ is finite, and each variation in the process requires time which is not smaller than the minimal and greater than zero, the class of variations forming a continuous series in $\{t^1, t^2\}$ is finite. Note that the time, required for each variation, is a fixed quantity. As we have already said above, it cannot vary.

A more restricted concept of rate can be obtained in the following manner: (1) we select the uniform variations of a given process (i.e., variations from a certain class B); (2) the interval $\{t^1, t^2\}$ is decomposed into uniform and equal parts belonging to a certain class C; (3) the rate of A is taken to be equal to the number of variations from the class B within the interval of time from the class C. Finally, the standard units of measurement are chosen. But all this does not affect the logical foundation of the matter: the rate of a process is the number of variations of this process during a certain time.

A special case of the process is provided by motion of a body (variation of a body's location in space). In this situation the above-mentioned class B is formed by displacements of equal length. There are standard units of distance. When the velocity (rate) of motion of a body is interpreted as the ratio of the distance covered by it to the time required, we implicitly perform the operation described above: the number m of equal consecutive displacements of a body over a time n is regarded as the magnitude of a distance covered by the body during this time.

The following consequences may be obtained from the definition of the rate. The rate of any process is finite. The following argument supports this statement: each variation of a process requires a time k which is greater than zero and is not smaller than the minimal time; it is either equal to or less than n, where n/k represents a finite quantity.

The rate of any process cannot exceed that in the case for which any variation in the process consumes the minimal time, and the variations of the process form a continuous series. Then the number m will be the maximal. It is also evident that if $m = 1$, it will correspond to the lower limit of rate – the rate of a process cannot be less than $1/n$.

This result can be expressed as follows:

$$(\mathrm{T}^1) \qquad (\forall x)(\forall \alpha)(\exists \beta)(\beta \geqslant \alpha),$$

where x is a variable for processes, whereas α and β are variables for rates of x.

Some additional conclusions can be obtained as well, which we will present in the form for a special case of displacement (these can be generalized for any process through mere paraphrase).

Consider the following statement: for any velocity (rate) of motion of bodies (in general, of a process) x there is (a possibility of) a velocity y of motion of bodies such that $y > x$. What is the meaning of this statement if we compare it with the definition of velocity (rate)? It means that if a body A covers a distance s during a time t^1, then there is a possibility of such a motion of a body B (B, in particular, may coincide with A) that the same distance s is covered by the body B during a time t^2, with $t^2 < t^1$. And so on, without end. Therefore, the time, spent in covering s, may be infinitesimal. But this result contradicts the assertion according to which the time spent on any variation cannot be less than the minimal. Thus, when time reaches the minimal value, the velocity (rate) of motion of bodies will be maximal. Our assertion at the beginning of the paragraph is not acceptable. The following statement is correct: there exists a velocity (rate) of motion such that is greater than any arbitrary velocity of motion; in other words, there is a maximal velocity. If a and b are individual variables and q is velocity, we obtain:

$$(\mathrm{T}^2) \qquad \vdash (\exists a)(\forall b)(a \geqslant qb).$$

This assertion is a consequence of the statement of the existence of the minimal time, in other words,

$$\vdash (\exists \alpha)(\forall \beta)(l\alpha \leqslant l\beta) \rightarrow (\exists a)(\forall b)(a \geqslant qb).$$

Formula (T^2) results from comparing the rates of two arbitrary processes x and y: a finite time n is chosen; the number of variations of any of these processes during this time cannot exceed the number m of variations such that each of them requires the minimal time; thus, whatever the rate of the other process, it cannot exceed m/n.

We may also present the assertion of the existence of a velocity y which exceeds velocity x in such a form that for velocity y the body will move further than for velocity x. This interpretation, however, can be reduced to the one discussed earlier: consider an arbitrary interval of the distance covered by the body moving with a velocity x and compare it with the same interval in the case of velocity y; we conclude that the same distance can be covered during different times, a time decrease below the minimal value being impossible in this situation too.

Rate (velocity) is always relative in the following sense: rate is always a rate of a certain process, and the process always represents a series, ordered with respect to a certain α.

Imagine the following situation: an observer A records a certain event x at a time t^1 in a spatial domain s^1, while an observer B records a certain event y at a time t^2 after t^1 in a spatial domain s^2 such that s^1 and s^2 represent different locations (in the strong sense). Assume that t is a temporal interval between t^1 and t^2, and s is a distance between s^1 and s^2 (with respect to a certain α). Forming a ratio of s to t, we shall not yet obtain a value of certain velocity, i.e., a rate of a certain empirical process, as we have not specified exactly which process is meant. In the abstract sense one may assume that some process W has occurred in this situation. And also in the abstract sense we may assume any value for s and any value for t (of course, positive), so that we conclude that the rate of W has no limits (may be infinitely great as well as infinitesimal). The empirical aspect of the matter, however, is quite different.

From the empirical point of view, either the displacement of a body a from s^1 (a state x) into s^2 (a state y) occurred, or the observer moved from s^1 and s^2, or the observer recorded x in s^1 and then y in s^2, or the observer A communicated to the observer B a message to the effect that she recorded x in s^1 at t^1 or an observer C watches the behavior of A and B, etc. And in all cases some empirical process should be specified for which the above statements are valid: that of the impossibility of temporal intervals and spatial distances that are smaller than the minimal, and, as a consequence, the assertion of the impossibility of motion with a velocity exceeding the maximal one or below the minimal.

In the abstract situation under consideration, we should assume, in addition to everything else, the existence of such a process z that x and y represent elements of this process, y exceeds x in temporal order, and the time spend on every element of z not less than the minimal. And then we can formulate the following result. If for this z a conclusion is reached, 'The rate of z in a temporal interval t is greater than the maximal', then either x and y do not represent elements of the same process z or some errors crept into the argument employed to derive the above conclusion.

We would like to stress the fact that we are dealing not with a certain concrete value of rate, but rather with such expressions as 'the rate is maximal', 'the rate is minimal', etc. and with assertions containing these expressions 'the rate of z is greater than the maximal', 'the rate of z is smaller than the minimal', etc.

We can obtain still another interesting result for velocity. We shall say that a body A moves continuously in a temporal interval $\{t^1, t^2, \alpha\}$ passing a series of points, ordered with respect to β, if and only if the following is valid: for any pair of time moments t_1 and t_2 such that $t_1 \| \alpha\, t_2, t^1 \geqslant \alpha t_1$ and $t_2 \leqslant \alpha t^2$ the body A occupies a point s^1 at t_1 and a point s^2 and t_2, with $s^2 > \beta s^1$. According to the definition, during the minimal time the body A covers a distance which is not less than the minimal. And therefore, the distance, covered by the body A during the time interval $\{t^1, t^2, \alpha\}$ is not less than the minimal. If we define the minimal velocity of the motion of a body A during the time interval $\{t^1, t^2, \alpha\}$ as the ratio of the minimal distance covered by the body to the time spent on it, it becomes obvious that if the body A moves in a continuous fashion, its velocity is not less than the minimal.

24. Zeno's Paradoxes

There is an opinion that Zeno's paradoxes have not been resolved up to the present time. And this is indeed so, since, in resolving 'paradoxes of Zeno', various authors usually discuss some of their own problems instead of the real paradoxes, which are either trivial or do not exist at all.

It is not possible to analyze these paradoxes here in detail. Let us make just a brief remark concerning the paradox of 'Achilles and the tortoise'. This paradox is formulated as follows: Achilles and the tortoise move in the same direction; Achilles lags behind the tortoise by a distance s^1; the velocity of Achilles' motion exceeds that of the tortoise; when Achilles covers a distance s^1, the tortoise will cover in the same time a distance s^2, and Achilles will have to pass s^2; when this has been done, the tortoise will cover in this time a distance s^3, and so on, without end, for any s^i; hence the conclusion that Achilles will never catch up with the tortoise.

This conclusion, however, is not acceptable: it does not follow logically in any way from the above premisses. And in the attempts to resolve the paradox, no references to mathematics and physics make sense, since this conclusion is obtained not on the basis of the logical rules, but from certain psychological and linguistic associations. In order to obtain the result 'Achilles shall never catch up with the tortoise', the conditions planned should be reformulated as follows: until the tortoise has covered a distance s^i, Achilles must not cover a distance, exceeding s^{i-1}. This condition can be visualized in this way: Achilles is connected to the tortoise by a rod which can contract to an arbitrary length but never disappears (never becomes zero). Under this condition, indeed, Achilles will never catch up with the tortoise. But this is true only for an abstract process. As far as an empirical process is concerned, even under the above conditions there comes a moment when Achilles reaches contact with the tortoise – the interval between them reaches the minimal value (and 'the rod' connecting them becomes such that no individual can be placed between them.) Let us also recall the fact that the contraction of the interval between Achilles and the tortoise occurs in 'jumps', and if nothing prevents this, the interval can be contracted to its minimal value.

25. Quanta of Space, Time and Motion

In the previous sections (§21) we have already stated that the 'motion' of a towards b in the case of an interval $\{a, b, \alpha\}$ or the choice of an interval $\{c, d, \alpha\}$ which is shorter than $\{a, b, \alpha\}$ occurs in jumps. Generally speaking, this means that if two intervals x and y have different dimensions, the difference in their lengths cannot be less than a certain minimal quantity, i.e., the minimal length of a certain individual. Accordingly, if two individuals x and y have different dimensions, the difference between them cannot be less than the minimal.

Thus, if all bodies and distances are ordered by length, a dimensional hierarchy results such that for any pair x and y of dimension values we shall have either $x = y$ or one of x and y exceeds the other by the minimal length. The argument for the duration of the existence of individuals and temporal intervals is similar.

Thus, in a sense space and time show a quantum nature; this does not mean, however, that they literally consist of spatial and temporal quanta, but rather it means that in the final analysis objects differ in their spatial and temporal dimensions by discrete quantities, or quanta.

26. The Relativity of Motion

Speaking of the relativity of motion, one has to differentiate between two aspects: (1) recording spatial positions of a body with respect to a certain method of establishing their spatial positions; (2) comparison of the motions of various bodies. In the first case, the method of establishing the spatial position of bodies implies the existence of a certain object with respect to which the position of the body is fixed, as well as a series of positions in time, allowing us to speak of translation of a body. There may be different methods for the description of motion. For example, the motion of a body A counterclockwise in a circle may be fixed with respect to the centre of the circle B or clockwise with respect to a body C, placed on the circle. In the second case of motion of each body we should specify that method of establishing the spatial order with respect to which it is fixed, regardless of the motion of other bodies. If this requirement is not fulfilled, it makes no sense to speak of the comparative motion of different bodies. Thus, in order to construct a logically correct statement concerning how a certain body A moves with respect to another body B, one has to compare the motion of A with respect to a certain method x of establishing spatial order, and the motion of B with respect to another method y of establishing the spatial order.

Consider an expression 'A body A moves toward a body B'. It appears obvious that from this expression a statement 'A body B moves towards a body A' follows logically. However, this appearance is misleading. In order to speak more precisely about this, we must consider an expression 'A body A moves towards a body B with respect to x', where x fixes a certain method of establishing the spatial order. Then we may discover that our expression is used with at least three different meanings: (1) as a replacement for a set of expressions 'A body B is at rest with respect to x', 'A body A moves with respect to x', 'The distance between A and B with respect to x decreases'; (2) as a replacement for a set of expressions 'A body A moves with respect to x' and 'The distance between A and B with respect to x decreases'; (3) as a replacement for an expression 'The distance between A and B with respect to x decreases'. In the first case, from the expression 'A body A moves towards a body B with respect to x' logically follows the expression 'A body B does not move towards a body A with respect to x', and, therefore, does not follow the expression 'A body B moves towards a body A with respect to x'. In the second case, from the expression 'A body A moves towards a body B with respect to x' the expression 'A body B moves towards a body A with respect to x' does *not* follow logically, although it is not excluded (as in the first case). It is only in the third case that from the expression 'A body A moves towards a body B with respect to x' we can obtain the expression 'A body B moves towards a body A with respect to x'.

Consider now the question of the motion of one body around another. Here also two cases must be distinguished: (1) the motion of one body around the other when the latter is a reference point for the spatial positions of bodies (i.e., it constitutes an element of the method of establishing the spatial position of bodies); (2) the motion of one body around the other when the motion of bodies is fixed with respect to one another independently of any method, and none of the bodies is taken into account by this method.

Consider the statement: 'The Sun moves around the Earth' (or 'The Earth moves around the Sun'. This statement does not belong to the type 'A moves around B with respect to *x*', although it appears to do so at the first glance. In reality in this case the Earth plays the role of a reference point for fixing the spatial position of the Sun (or, conversely, the Sun is the reference point for fixing the spatial position of the Earth). The method of recording the spatial position of the Sun with respect to the Earth is *x* (it is well known, and will not be discussed), while that of recording the spatial position of the Earth with respect to the Sun is *y*. In this situation we deal with an assertion 'The Sun moves in such and such a way with respect to *x*' or, conversely, 'The Earth moves in such and such a way with respect *y*'. There is no logical relation between these whatsoever; in other words, from one of them neither the other one nor its negation follows. There is no logical contradiction between them at all.

A different picture appears when the comparative motion of two bodies A and B is considered under the condition that the motion of each of them is recorded regardless of fixing the motion of the other with respect to a certain method *x*. In this situation, obviously, neither A nor B represent an element of *x*. As far as the Sun and Earth are concerned, we should specify a certain method *x* that does not mention the Sun and the Earth as the reference points for registering the spatial position of bodies.

Consider, further, an expression: 'A body A moves around a body B with respect to *x*'. Certainly, it implies a temporal interval. But in order to simplify the record, we shall omit reference to it, since it is the same for all expressions. Various explications of this expression are possible, in particular, the following one.

Assume that $\alpha^1, \alpha^2, \ldots, \alpha^n$ are consecutive positions of a body in space with respect to x at the time moments, respectively, $t^1, t^2, \ldots, t^n$ in a temporal interval t. The spatial domain of location of A with respect to x in t will be called an area in space such that all $\alpha^1, \alpha^2, \ldots, \alpha^n$ are either the boundary points of this spatial domain or inside it (with respect to x). The definition for B (with $\beta^1, \beta^2, \ldots, \beta^n$ being the positions of B with respect to x at time moments $t^1, t^2, \ldots, t^n$ in t) are similar. Assume that V is the spatial domain of location of A with respect to x in t, and W that of location of B with respect to the same x, at the same time t. We shall say that a body A moves around a body B with respect to x in t if and only if W is inside V, and for any α^i and any β^i there is an α^k such that β^j is between α^i and α^k with respect to *x*. In the case of other explications the defining part of a given expression should in one or another way be obtained as a consequence. A special case of the above explication: a body A moves around a body B with respect to *x* if and only if the trajectory of A with respect to *x* represents a line such that B is inside it with respect to *x*.

According to the definition of the expression 'is inside', the following consequence results from the above definition: if W is inside V, then V is not inside W with respect to the same *x* at the same time. And therefore, if A moves around B with respect to *x* in *t*, then B does not move around A with respect to the same *x* at the same time *t*.

We do not exclude a possibility that B may move around A with respect to another method, say, *y* (the fact that it is possible at another time is self-evident). Let us now perform this operation: define an expression 'A body *a* moves around a body *b* with respect to *z*', where *a* and *b* represent variables for terms of arbitrary bodies. It will be defined in the following manner: a body *a* moves around a body *b* with respect to *z* if and only if there is a method *c* of recording the spatial position of bodies such that *a* moves around *b* with respect to *c* (where *c* is a variable for the methods of establishing the spatial position of bodies). As A and B represent individuals from the domain of values of *a* and *b*, and *x* and *y* – from the domain of values of *c*, we obtain the result that (1) A moves around B with respect to *x*, therefore, A moves around B with respect to *z*; (2) B moves around A with respect to *y*, therefore, B moves around A with respect to *z*. Thus, A and B move around each other with respect to the same *z*.

However, according to the rules of definitions of linguistic expressions, we have defined not the expression *z* by itself, but an expression 'A body *a* moves around a body *b* with respect to *z*' as a whole. And *z* is entirely meaningless outside of this whole. The expression *z* does not represent a generalization of *x* and *y*, although it appears to. Thus, the expressions 'A body A moves around B with respect to *z*' and 'A body B moves around A with respect to *z*' cannot be interpreted in the sense that A and B move around each other with respect to the same method of establishing the spatial order. According to the first of them, A moves around B with respect to a certain method of establishing spatial order, whereas according to the second, B moves around A with respect to a certain method. But this does not mean that the method is the same in both cases.

Further, let us adopt a Definition D^1: A body *a* moves continuously away from a body *b* with respect to α in a temporal interval *t* with a velocity *v*, if and only if for any two time moments *x* and *y* in *t* the following holds: if *y* is after *x*, then $[ab\alpha y] = [ab\alpha x] + v[xy]$, where $[ab\alpha y]$ and $[ab\alpha x]$ represent the length of the interval between *a* and *b* with respect to α at *y* and, respectively, at *x*; $[xy]$ is the duration of the time interval between *x* and *y*.

From D^1 follows assertion X^1: if $[ab\alpha z] = [ba\alpha z]$, where *z* is any of the time moments in *t*, and a moves away from *b* with respect to α in *t* with a velocity *v*, then *b* moves away from *a* with respect to α in *t* with the same velocity *v*. But we cannot derive from D^1 an assertion X^2: if *a* moves away from *b* with respect to α in *t* with a velocity *v*, then *b* moves away from *a* with respect to α with a velocity *v*. The assertion X^2 is extralogical.

If we accept an assertion X^3: $[ab\alpha z] = [ba\alpha z]$, then X^2 will represent a logical consequence of X^1 and X^3. The assertion X^3 is also extralogical. It is not always true. For instance, it is not always true in the case when the distance between *a* and *b* is determined by the motion along a circle clockwise, the counting starting with an object which is called the first.

From D^1, further, an assertion Y^1 can be derived: if $[ab\alpha z] + [bc\alpha z] = [ac\alpha z]$, *a* moves away from *b* with respect to α in *t* with a velocity v^1, and *b* moves away from *c* with respect to α in *t* with a velocity v^2, then *a* moves from *c* with respect to α in *t* with a velocity $v^1 + v^2$. But from D^1 we cannot obtain an assertion Y^2: if *a* moves away from *b* with respect to α in *t* with a velocity v^1, and *b* moves away from *c* with respect to α in *t* with a velocity v^2, then *a* moves away from *c* with respect to α in *t* with a velocity $v^1 + v^2$. The assertion Y^2 is extralogical.

If we adopt an assertion Y^3: $[ab\alpha z] + [bc\alpha z] = [ac\alpha z]$, then Y^2 will represent a logical consequence of Y^1 and Y^3, in other words, the rule of summation of velocities Y^2 will represent a consequence of the rule of summation of distances Y^3. The assertion Y^3 is also extralogical and is not always true. For example, it does not hold for points *a*, *c*, *b* on a circle, located in the above order, if the distances are measured by the method described previously.

Let us adopt, further, a definition D^2: A body *a* moves continuously away from the location of a body *b* with respect to α in a temporal interval *t* with a velocity *v* if and only if for any time moments *x* and *y* in the interval *t* and for any individual *c* the following is valid: if *c* is identical to *b* in order with respect to α in *x*, and if *c* is at rest with respect to α, then $[ac\alpha y] = [ab\alpha x] + v[xy]$. Obviously, the case of D^2 is more general than that of D^1.

Now we may consider such cases as (1) a moves away from the location of b with a velocity v^1, b moves away from the location of c with a velocity v^2; (2) a moves away from b with a velocity v^1, b moves away from the location of c with a velocity v^2; (3) a moves away from the location of b with a velocity v^1, b moves away from the location of c with a velocity v^2. In particular, it may be established in a purely logical way that in case (1) body a moves away from b with a velocity $v^1 - v^2$. Obviously, it is logically impossible for v^2 to exceed v^1.

The cases just considered seem to be very characteristic. The assertions X^2, X^3, Y^2, Y^3 seem to be logically obvious, while the statement of the existence of the maximal velocity appears as a purely physical assumption. In reality, the former are not logically true, while the latter represent a logical consequence of the definition of linguistic expressions in the frame of logic. This discrepancy between the apparent and the actual status of statements does not constitute an exception to the rule; this we have seen on several occasions.

There are logical boundaries for adoption of assumptions such as X^2, X^3, Y^2, Y^3. Thus, if the acceptance of Y^2 or Y^3 results in an assertion Z: 'The velocity $v^1 + v^2$ is greater than the maximal', then we have at least one of the three possibilities: (1) it is not true that a moves away from b with a velocity v^1; (2) it is not true that b moves away from c with a velocity v^2; (3) it is not true that assertions Y^2 and Y^3 hold for given bodies. To put it briefly, the physical fact, specified in Z, is logically impossible and therefore impossible from the physical point of view.

27. On the Existence and Motion of Clusters

At least four different existence predicates E^1, E^2, E^3, E^4 are possible in the case of clusters of empirical bodies. These are defined implicity by the following statements (where A represents the individual cluster term, a is a variable for the terms of empirical individuals, b is a variable for the specified class terms, c is a variable for cluster terms):

(1) $E^1(A) \dashv\vdash (\forall a)E(a\downarrow(a \in A))$,

(2) $\neg E^1(A) \dashv\vdash (\exists a)\neg E(a\downarrow(a \in A))$,

(3) $E^2(A) \dashv\vdash (\forall b)(\exists a)E(a\downarrow((a \in b) \wedge (a \in A)))$,

(4) $\neg E^2(A) \dashv\vdash (\exists b)(\forall a)\neg E(a\downarrow((a \in b) \wedge (a \in A)))$,

(5) $E^3(A) \dashv\vdash (\exists c)(E^1(c) \wedge (c \subset A))$,

(6) $\neg E^3(A) \dashv\vdash (\forall c)(\neg E^1(c) \wedge (c \subset A))$,

(7) $E^4(A) \dashv\vdash (\exists c)(E^2(c) \wedge (c \subset A))$,

(8) $\neg E^4(A) \dashv\vdash (\forall c)(\neg E^2(c) \wedge (c \subset A))$.

The duration of existence of A in the sense of E^1 is equal to the minimal duration among those of the existence of bodies constituting A. According to (3) and (4), it is assumed that A exists in the sense E^2, although some of the bodies, constituting it, are replaced by bodies of the same kind. From this point of view, the duration of the existence of a cluster may exceed that of the existence of its elements. As follows from (5)–(8), the duration of existence of a cluster is determined by the duration of existence of one of its parts. For example, a person exists in the sense of (3) and (4), although some cells of her body are constantly replaced by new ones, and she continues to exist in the sense of (5)–(8) after the amputation of a leg.

If the duration of existence of elements of a cluster is close to minimal, the time of existence of the cluster will coincide with that of its elements. Thus, one cannot apply the concepts of genesis, evolution, development, etc. to this type of cluster, since they appear and become destroyed 'at once', together with all their elements, independently of connections and relations between the latter. Also evident is that it makes no sense to talk of causal and other empirical relations between the elements of such clusters.

Assume, further, that b^1 represents the position of a body a at a time t^1 with respect to α, and b^2 – the position of a at a time t^2 after t^1 with respect to α. Speaking of distance at which the body a moved with respect to α in a temporal interval $\{t^1, t^2\}$, we shall have in mind the spatial intervals $\{b^1, b^2, \alpha\}$.

If one talks about a cluster of bodies, it is necessary to formulate a special agreement concerning how to define the distance of the cluster's displacement. The following definition is appropriate in this case; the distance through which a cluster A moved with respect to α during a time t is the minimal quantity among those distances at which all bodies constituting A moved with respect to α during the time t. Hence a conclusion that the velocity of A's motion equals the minimal value among the velocities of motion of bodies constituting A. If the dimensions of, and the distances between, the elements of A are close to the minimal values, the velocity of motion of A equals that of its elements. When it is possible to distinguish between the velocities of bodies' motion, agreements are adopted in practice to define the distance of A's motion and its velocity only by means of a certain cluster B such that $B \subset A$ (for instance, in considering a motion of an army division, individual soldiers and even whole battalions, lagging behind, are not taken into account).

28. The Beam

Imagine a body A and a body B at a certain distance from it. A variation α occurs in A or because of it. As a consequence of this, after some time a variation β occurs in B. What happens between A and B at this time? From the logical point of view, there are the following possibilities: (1) a certain body moves from A towards B (α representing the beginning of this motion, β, its end, the encounter with B); (2) a wave process takes place in the medium between A and B; (3) a combination of (1) and (2). But one can imagine at least one more possibility, different from (1)–(3). And it emerged in one or another form in the explanation of the phenomena of the microworld; this is an illustration of our thesis according to which, from a linguistic point of view, research work undertakes to realize just logically imaginable possibilities, without going beyond their limits.

Consider empirical bodies or clusters of empirical bodies a^i ($i = 1, 2, \ldots, n$) belonging to a class a (or all of them constituting a). The body a^1 moves at a distance x^1 with respect to α (say, in the direction β); t_1^1 represents the time moment of the beginning of motion of a^1, and t_1^2 the time moment of the end of the motion of a^1; the distance x^1 is finite, the time interval $\{t_1^1, t_1^2\}$ is finite; at t_1^2 body a^1 reaches body a^2, comes in contact and interacts with it; $\{t_2^0, t_2^1\}$ represents an interval of time during which interaction takes place; as a result of interaction between body a^1 and body a^2, a^2 starts moving in the same direction β (with respect to the same method α); body a^2 starts its motion at t_2^1, terminates it at t_2^2, moves over a distance x^2, comes in contact with a^3, interacts with it, spends a time $\{t_3^0, t_3^1\}$ on interactions, and so on for all bodies a^i and a^{i+1} from a given class a. These translations and interactions from a continuous series, ordered in time.

Let us adopt a definition: the ordered series of displacements and interactions of bodies (clusters of bodies) will be called an empirical beam. The position of a^1 at t_1^1 will be interpreted as the beginning of the beam, and the position of a^n at t_n^2 as its end. The length of the interval between the beginning and the end of a beam is the length of the beam. The interval $\{t_1^1, t_n^2\}$ will be interpreted as the time of existence of an empirical beam. A simple case of a beam is a single displacement a^i. The elementary beam is one in which all a^i's represent empirical bodies, not decomposable into parts (i.e., the dimensions of which approach the minimal). The possibility discussed at the beginning of this section, corresponds to the beam.

As the length of a beam x and the time of its existence are finite, the quantity x/y (the mathematical or abstract velocity of beam propagation) is finite.

There may be different meanings to the notion of the empirical velocity of beam propagation, in particular, the following: (1) the number of uniform consecutive translations in $\{t_1^1, t_n^2\}$ is interpreted as the beam velocity: (2) a certain standard process is chosen in $\{t_1^1, t_n^2\}$ and its rate is regarded as the velocity of beam propagation; (3) $a^1, a^2, \ldots, a^n$ are considered as the same a, and their translations as a translation of this a; the velocity of this translation is interpreted as the beam velocity.

If the bodies a^i are not differentiated in any way, the method of (3) may in general turn into that of understanding a given complex phenomenon. Then the sequence of translations of various bodies belonging to the same class is perceived as a single body's translation. If the dimensions of bodies and the time intervals approach the minimal, this creates the most favorable conditions for that kind of illusion. If it is actually impossible to establish that a^i and a^{i+1} represent different bodies, no error can be found in the interpretation of the beam as a translation of a single body a. When a^i represent clusters of bodies, the dimensions of and the distances between which are close to the minimal, then under the above conditions we can say, beside everything else, that a^i's are perceived not as clusters of bodies but as a single elementary body. One can add to this a combination of two and more beams into one complex phenomenon. Thus, from the logical point of view it is not surprising at all that such a phenomenon will reveal properties that appear unusual when compared with those exhibited in the cases of explicit motion of a single body.

We would like to point out the fact that the definition of the beam does not require the definition of wave or particle (see below). This is a different approach to empirical bodies and their motion. The corresponding phenomena observed can be described in different terminology, and it is possible that these descriptions will be in some sense equivalent.

29. The Universe as a Whole

The world (the Universe) represents a cluster of empirical individuals, including all empirical individuals; in other words, if x is a variable for empirical individuals, then

$$\vdash (\forall x)(x \in Universe).$$

It follows from the definition that the *Universe* is an empirical individual.

It also follows from the definition that the *Universe* is unique, i.e.,

$$\vdash (\forall x)(\forall y)((x \rightharpoondown \mathit{Universe}) \wedge (y \rightharpoondown \mathit{Universe})) \rightarrow (x \rightleftharpoons y)),$$

where x and y are individual variables (to put it differently, if x is the *Universe* and y is the *Universe*, then the terms x and y have identical significance or x and y represent the same individual).

From the definition we also obtain

$$\vdash \mathrm{E}(\mathit{Universe}) \equiv (\exists x)\mathrm{E}(x),$$
$$\vdash \mathrm{E}(\mathit{Universe}) \equiv (\forall x) \neg \mathrm{E}(x),$$

i.e., the Universe exists if and only if at least one empirical individual does exist, and the Universe does not exist if and only if no empirical individual exists.

The Universe represents a cluster of individuals, and one can apply predicates to it (as to any cluster), which are defined for clusters. But we must be careful here, because of the specific feature of the definition of this cluster and the ambiguity of linguistic expressions containing the words 'emerges', 'infinite'. etc.

Consider an expression: 'The Universe did not emerge in time' (or 'The Universe has no beginning in time', 'The Universe is eternal in the past', etc.) It is ambiguous. First, it can be understood as the negation of the assertion 'The Universe did emerge in time', which is an abbreviation for

$$\neg \mathrm{E}(\mathit{Universe}) \Rightarrow \mathrm{E}(\mathit{Universe}),$$

in which the Universe is considered merely as an empirical individual. However, in order either to accept or to disprove such an assertion, it is necessary to have a reference point for time, a certain empirical event a, to have time markings, i.e. empirical events $b^1, b^2, \ldots$, to be able to observe a state $\downarrow \neg$ E(Universe, with individuals a, $b^1, b^2, \ldots$ not included in the Universe. And according to the definition, this is impossible. If, on the other hand, $a, b^1, b^2, \ldots$ are included in the Universe, it follows from the definitions and from E(a), E(b^1), E(b^2), ... that E(*Universe*), in other words, a state $\downarrow \neg$ E(*Universe*) is in principle impossible to observe. Thus, in this sense our assertion is indefinite, i.e., it is not true that 'The Universe emerged', and it is not true that 'The Universe did not emerge'.

To sum it up, we have arrived at something opposite to Kant's antinomies, namely the negation of both opposed statements. But in this case it does not result in contradiction, since there is a third possibility, which is uncertainty.

The second meaning of the above assertion is as follows: The Universe is regarded as a process, or, to put it differently, as a series of various states in time. Then our assertion means that this series has no starting element, i.e.,

$$(\forall x)(\exists y)(x > \alpha y),$$

where x and y are variables for states of the Universe, and α is a method of their ordering in time with respect to a certain event from the Universe (it is not forbidden here).

Leaving aside the question whether or not this assertion can be verified in practice, we can say that from a purely logical point of view it is meaningful and is not in conflict with the adopted definitions. In particular, from the fact that the Universe has no beginning in time (with the second meaning) it does not follow that its existence is impossible.

Similarly, one can neither accept nor reject the assertion to the effect that the Universe does not move in space, if the Universe is regarded simply as an individual, since the bodies that are not included in the Universe and with respect to which the motion or the rest state of the Universe is recorded are impossible.

The rules of predication for subjects with expressions 'The Universe in space', 'The Universe in time' are identical to the rules for subjects with expressions 'Space' and 'Time'. Thus, these expressions are virtually used as synonyms. They can be explicated in different ways, in particular, as follows.

The Universe is actually (potentially) finite in space with respect to a class of methods of establishing a spatial order A if and only if $(\forall x)(\exists a)(\forall b)((x \in A) \rightarrow (a \geqslant xb))$, where a and b represent variables for empirical individuals, and x a variable for the methods of establishing a spatial order (respectively, $(\forall x)(\mathrm{M}a)(\forall b)((x \in \mathrm{A}) \rightarrow (a \geqslant xb))$. The Universe is actually (potentially) infinite in space with respect to A if and only if $(\forall x)(\forall a)(\exists b)((x \in \mathrm{A}) \rightarrow (b > xa))$, (respectively, $(\forall x)(\forall a)(\mathrm{M}b)((x \in \mathrm{A}) \rightarrow \rightarrow (b > xa)))$. From a logical point of view it is not impossible that for a certain A the Universe is neither finite nor infinite. The Universe is finite (infinite) in time in the past with respect to β if and only if $(\exists x)(\forall y)(x \leqslant \beta y)$ (respectively, $(\forall x)(\exists y)(y < \beta x)$), where β is a method of establishing the temporal order, and x and y represent variables for empirical measurements. The Universe is finite (infinite) in time in the future with respect to β if and only if $(\mathrm{M}x)(\forall y)(y \leqslant \beta x)$ (respectively, $(\forall x)(\mathrm{M}y)(y > \beta x)$). The expressions 'The Universe is potentially finite (infinite) in the past' and 'The

Universe is actually finite (infinite) in the future' are meaningless.

The common feature of all possible explications of the assertions under consideration is that they are neither logically true nor logically false, that they are independent of logic and logically realizable. It is only their contradictory combinations that are logically non-realizable.

As follows from the above definitions, the assertion 'The Universe is not finite and is not infinite in space with respect to a certain A' is logically realizable.

The following definitions are equivalent to those given above: (1) the Universe is existentially (potentially) finite in space with respect to A if and only if there exists (is possible) a spatial structure B such that any other spatial structure C represents a proper substructure of B with respect to A; (2) the Universe is existentially (potentially) infinite in space with respect to A if and only if for any spatial structure B there exists (is possible) such a spatial structure C that B represents a proper substructure of C with respect to A. Equivalent to these definitions are those obtained from (1) and (2) through substitution of the following expressions: (1) that any empirical individual is inside B with respect to A or represents a boundary point of B; (2) there exists (is possible) an empirical individual which is not inside B with respect to A and does not represent a boundary point of B.

The infinity and the finiteness of the Universe in space can be defined through motion of a body. But these definitions will be more narrow. Besides, they require a whole system of terms which we do not have yet at our disposal, in other words, they are logically more complex.

The defining part of the notions of the finiteness and the infinity of space and time is the same as that of the finiteness and infinity of the Universe in space and time. And in this sense the expressions 'Space (time) is finite (infinite)' are identical, respectively, to the expressions 'The Universe is finite (infinite) in space (in time)'.

30. Empirical Geometry

We shall employ the following abbreviations: (1) EP, empirical point; (2) EL, empirical line; (3) ES, empirical surface; (4) EB, empirical body.

D1. Definition of an empirical point; EP is an empirical individual, characterized by minimal spatial dimensions with respect to any method of establishing a spatial order.

Every EP has dimensions. The magnitude of these dimensions is greater than zero but it is the minimal (the smallest). It is a question of practical agreement which objects exactly should be considered as EP's; the nature of this agreement depends on extralogical factors (in the same way as the very assumption of the existence of material points in physics does not contain an indication as to exactly which physical bodies may be regarded as material points).

In D1 we have defined the term 'EP'. And although in the defining part there is a reference to the methods of establishing a spatial order, as far as the application of the term 'EP' itself is concerned, this is of no importance: the definition is indifferent as to the concrete choice of method. And in this sense one can say that EP is zero-dimensional.

In what follows the symbols α, β, γ, δ will be used as notations for certain classes of methods of establishing a spatial order. In exactly the same way we shall assume the existence of arbitrary classes of this kind (i.e. it does not matter which ones). But in this case indifference with respect to the concrete nature of these classes will be of another character, as these symbols will appear both in the defining and in the defined parts of the definitions.

D2. Definition of an empirical line: (1) EP is EL with respect to α; (2) a continuous series of EP's, ordered with respect to α, is EL with respect to α; (3) something is EL with respect to α only by virtue of (1) and (2).

Every EP is EL with respect to any method of establishing the spatial order, but not every EL represents EP; this is obvious, for instance, in the case of an ordered series of two and more EP's. As follows from D2, if two EP's on a given EL are adjacent, they are in contact.

We would like to point out an important circumstance: in D1 we define the expression EP, and in D2 the expression 'EL with respect to α'. These additional words 'with respect to α' appears here as a result of the fact that in (2) of definition D2 we have mentioned an ordered series of EP's, which implies a specified class of the methods of establishing order. And in this sense EL is one-dimensional.

D3. The length of EL is the number of EP's belonging to it.

D4. EL *a* represents a segment of EL *b* if and only if every EP belonging to *a* is an EP belonging to B.

D5. The distance between EP's of a and b with respect to α is the number of EP's between a and b on EL with respect to α, which contains a and b.

D6. The shortest EL between EP's of a and EP's of b is EL between a and b such that is not longer than any other EL among those possible between a and b.

D7. EL a represents a straight EL if and only if for any two of its non-adjacent points a and b the following is valid: the segment of EL A between a and b represents the shortest EL between a and b.

D8. The shortest distance between two EP's is the length of the shortest EL between them or is minimal (EP's are in contact).

D9. EL A adjoins EL B with respect to β if and only if every point of A is in contact with at least one point of B with respect to β.

D10. EL's A and B represent adjoining EL's if and only if A adjoins B or B adjoins A (or both).

D11. Two EL's are different if and only if there is at least one EP which is not common for two of them.

D12. Definition of an empirical surface: (1) EL with respect to α is ES with respect to a pair of classes (α, β) of methods of establishing spatial order; (2) a continuous series of adjoining EL's, ordered with respect to β, with each of EL's being EL with respect to α, is ES with respect to (α, β); (3) something is ES with respect to (α, β) only by virtue of 1 and 2.

Every EL is ES, but not vice versa. In the definition D12 it is not the expression 'ES' which is being defined, but the expression 'ES with respect to (α, β)'. The addition of β is associated with the fact that not only the ordering of EP's in EL is implied, but also the ordering of EL's in ES. And in this sense ES is two-dimensional.

D13. ES A adjoins ES B with respect to γ if and only if every EP belonging to A is in contact with at least one EP belonging to B with respect to γ.

D14. Two ES's A and B adjoin each other with respect to γ if and only if A adjoins B or B adjoins A with respect to γ.

D15. EL belongs to ES if and only if all EP's of this EL represent EP's of this ES.

D16. The area of ES is the number of EP's belonging to it.

D17. The plane represents ES such that the shortest EL between any two of its EP's belongs to this ES.

D18. Definition of an empirical body: (1) ES with respect to (α, β) is EB with respect to a triplet of classes (α, β, γ) of methods of establishing the spatial order; (2) a continuous series of adjoining with respect to γ ES's with respect to (α, β), ordered with respect to γ, is EB with respect to (α, β, γ); (3) something is EB with respect to (α, β, γ) only by virtue of 1 and 2.

Thus, 'EB' is defined as 'EB with respect to (α, β, γ)', i.e., with respect to three classes of methods of establishing the spatial order. In this case γ is added because the series of ES's under consideration, ordered with respect to γ, consists of elements which are two-dimensional themselves. And in this sense EB is three-dimensional. Every ES is EB, but not vice versa.

D19. Two EB's adjoin each other with respect to δ if and only if one ES of one of them adjoin some ES of the other with respect to δ.

Consider now a continuous series of EB's with respect to (α, β, γ) adjoining with respect to δ, ordered with respect to δ. Since no limitations whatsoever have been imposed on the classes of methods of establishing order α, β, γ, we can use $\alpha^*, \beta^*, \gamma^*$ instead, which represent unifications, respectively, of each of α, β, γ with δ. Then the above series of EB's will be EB with respect to $(\alpha^*, \beta^*, \gamma^*)$, i.e., it will be EB, too, regardless of the number of the terms in this series. And four-dimensionality can be logically reduced to three-dimensionality. Similarly, five-dimensionality can be logically reduced (through four-dimensionality) to three-dimensionality, and the same can be said about six-, etc. (in general, $(n+1)$)-dimensionality; the latter reduces to three-dimensionality in the final analysis through n-dimensionality.

A similar method of reduction of EB to ES, ES to EL, EL to ET is impossible since, according to the definitions, not every EB is ES, not every ES is EL, not every EL is EP.

Our commonly held and vague belief in the fact that the Universe is three-dimensional in space is based on implicit agreements concerning the meaning of the above linguistic expressions. Their own agreements are perceived by people as the compulsory force of the Universe itself. The degree to which the words dominate the consciousness of people is much greater than we usually assume.

Consider two EL's A and B. Let EP *a* belong to A and B. Assume that EP *b* belongs to A, *a* and is in contact with it. Assume that EP *c* belongs to A, follows after *a* and is in contact with it. Thus, *a* is between *b* and *c*. Let us adopt a definition:

D20. EL's A and B intersect at EP *a* if and only if the above requirement is fulfilled, with *b* and *c* not belonging to B.

It can be demonstrated that EP's belonging to B and adjacent to *a* also do not belong to A.

D21. Two EL's intersect if and only if they intersect at a certain EP.

Assume, further, that *a* and *b* represent an arbitrary pair of EP's such that (1) both of them belong to EL A; (2) they are in contact; (3) *b* is superior in order to *a* (in other words, *b* follows *a*; a method of recording order is certainly implied). Similarly, let *c* and *d* be an arbitrary pair of EP's such that (1) both of them belong to EL B; (2) they are in contact; (3) *d* is superior in order to *c* (with respect to the same method).

D22. EL's A and B are parallel if and only if the shortest distance between *a* and *c* equals the shortest distance between *b* and *d*.

We have the following theorem about parallel EL's: if EL's A and B are parallel, they do not intersect.

Proof of the theorem of parallel EL's. Assume that A and B are parallel and intersect at EP a^1. Consider a^1 as EP belonging to A. Consider EP b^1 belonging to B, which is in contact with a^1. The shortest distance between a^1 and b^1 is minimal. Further, consider EP a^2 which belongs to A, follows a^1 and is in contact with it. Consider EP b^2 which belongs to B, follows b^1 and is in contact with it. The shortest distance between b^2 and a^2 is equal to that between b^1 and a^1, i.e., it is minimal. Therefore, b^2 and a^2 are in contact. But this means that with respect to a certain method of establishing order, b^2 is superior in order to a^2 and b^1 is superior in order to a^1. Consider now EP a_1 preceding a^1 on A, and b_1 preceding a^1 on B. The distance between them is not shorter than the minimal since they do not coincide (according to the definition of intersection). Therefore, there is a certain method of establishing order with respect to which a_1 is superior in order to b_1 or b_1 is superior in order to a_1. But this does not hold for a^1, since for any method of establishing order a^1 is not superior in order to a^1. Thus, A and B are not parallel, and our assumption is wrong.

The question of the relation between Euclidean and non-Euclidean geometries as well as that of the relation of both of them to empirical geometry is answered on the basis of the introduction of the notion of quasiparallel lines, which is different from that of the parallel ones. There is at least one case when quasiparallel lines are parallel, i.e. when they represent straight lines on a plane. Generally speaking, however, it does not follow from the fact that lines are quasiparallel that they are parallel, and, therefore, it does not follow that they do not intersect. We do not exclude the possibility of their intersection. In this connection, let us consider one simplest case (but it is the most characteristic from a philosophical point of view).

D23. EL A is a quasistraight line on ES W if and only if for any two of its non-adjacent points *a* and *b* the following is valid: the segment of EL A between *a* and *b* is shorter than any other segment of EL between *a* and *b*, belonging to this ES.

Suppose that A and B are two quasistraight lines on ES W with only one EP in common. Let this EP be *a*. Assume that *c* and *d* are arbitrary EP's belonging to B and having the property such that *a* is superior in order to *c*, and *d* is superior in order to *a*, in other words, that *a* lies between *c* and *d*. Suppose that *e* is an arbitrary EP belonging to A.

D24. We shall call A *a* perpendicular to B on ES W if and only if the following holds: if the segments of B between *c* and *a* and between *d* and *a* are equal, then the segments of quasistraight lines on ES W between *e* and *c* and between *e* and *d* are equal.

All other cases of definition of perpendicular can be reduced to this simplest one.

D25. Two EL's A and B on ES W will be called quasiparallel on ES W if and only if any perpendicular to one of them on ES W is a perpendicular to the other on the same ES W.

Assume that C^1 and C^2 are perpendiculars to quasistraight lines A and B on ES W. Assume that (a^1b^1) and (a^2b^2) are segments of these perpendiculars such that a^1 and a^2 represent EP's belonging to A, while b^1 and b^2 – those belonging to B. From the fact that A and B are quasiparallel on ES W it does not follow that segments (a^1b^1) and (a^2b^2) are equal. It also does not follow that the shortest distances between a^1 and b^1 and between a^2 and b^2 are equal. Even if we suppose that segments (a^1b^1) and (a^2b^2) are equal to each other, this does not mean that A and B do not intersect. Consider, for instance, parallel lines A and B on a plane W and bend the plane in such a way that A and B intersect at a point *a*. The surface is no longer a plane but it still is the surface W. The lines A and B will remain quasiparallel. The above segments will remain equal.

To put it briefly, if we perform a sufficiently rigorous logical explication of the linguistic expressions, we come across a multiple meaning of the word 'parallel'. When this multiple meaning is eliminated, the mysticism created in philosophy in connection with non-Euclidean geometry disappears. Empirical geometry represents a logical means for such explication of concepts.

D26. The empirical figure (EF): (1) EL is EF; (2) a cluster of pairwise intersecting EL's is EF; (3) something is EF only by virtue of 1 and 2.

Further definitions deal with specific types of EF's and their properties. We shall make just two remarks to provide some perspective for this issue.

Consider two straight lines A and B in the same plane. Assume that they intersect at one and only one point *a* (*a* is their only point in common). Let *a* be the starting point of both A and B. Such two lines A and B form EF which is called the angle between the straight lines on a plane. The magnitudes of these angles, up to a certain limit established through agreement, may be defined by a number $n^0 + n^1 + \cdots + n^m$, where n^0 is the number of points of A that are in contact with the points of B, n^1 is the number of points of A located at a distance of one point from the closest points of B, ..., n^m is the number of points of A located at a distance of *m* points from the closets points of B. The magnitudes of other angles may be defined as the sum of the above numbers.

In the geometry just described all finite lines are commensurable. For instance, the diagonal of a square always has a finite length that can be expressed by a positive integer (i.e., by a number of points that cannot be further divided into parts). Thus, this number is not $a\sqrt{2}$, where a represents the length of the side of the square. This happens only in mathematical (abstract) geometry. In our conception, this number is $a\alpha$ where $\alpha = f(a)$, and the type of f is chosen depending on circumstances in such a way that $a\alpha$ is a positive integer between a and $a\sqrt{2}$. If the number of points in a equals two, $\alpha = 1$. If the number of points in a is infinite, $\alpha = \sqrt{2}$. But such a square consisting of EL's does not exist empirically. There is nothing unnatural in this. If, for instance, we consider as EP's the elementary particles of physics, the lengths of EL's belonging to EF will be expressed through such giant numbers that the mathematical inaccuracies of empirical geometry will be of no practical significance.

Empirical geometry differs essentially from abstract (mathematical) theory. But they are not in conflict since the relation between them is governed by special rules for interpretation of abstract objects appearing in mathematical geometry. In particular, mathematical points on a mathematical line cannot be interpreted as EP's on EL, since a finite segment of the mathematical line contains an infinite number of points, while a finite segment of EL contains a finite number of EP's. Only the following interpretation is legitimate in this case: every point of a mathematical line can be set in correspondence with at least one point of a material line. It is possible that different mathematical points are set in correspondence with the same EP, and one mathematical point with two adjacent EP's.

31. Empirical Connections

Statements, whose subject-terms are empirical objects and which are neither logically true nor logically not-true, are empirical statements. If x is an empirical statement, an object, denoted by the term $\downarrow x$, is an empirical state. The connection between empirical states is an empirical connection.

Assume that x is a statement concerning the connection of states a and b, and y is that concerning the connection of a and c, and z the connection of c and b. The connection between a and b will be called a mediated connection in time (in space) with respect to α if and only if there is a state c such that

$$\vdash (y \wedge z \rightarrow x),$$

with c being between a and b with respect to α. The state c is a mediator of the connection between a and b. The definition of the unmediated connection can be obtained from the above definition by introducing negation in front of the words 'a state c'.

Obviously, if the temporal (spatial) interval between a and b equals the minimal with respect to α, the connection between a and b is unmediated in time (in space) with respect to α. Thus, the process of locating the mediators is not infinite.

The process of location mediators is also not infinite in the case of a variation of the method of establishing an order. Let the time between a and b be equal to t. Having located a mediator c between a and b, we can locate a mediator between c and a only under the condition that the time between them is greater than the minimal. This time is less than t. Since t is finite, and its decrease accompanying the location of mediators occurs in a discrete manner, continuing the process we shall arrive at an interval which cannot be further decreased.

There is in principle a limit to the process of location of mediators $c^1, \ldots, c^n$ between a and b such that $a, c^1, \ldots, c^n, b$ form a continuous series with respect to α.

32. The Predicates of Tendencies

We frequently come across predicates in linguistic practice, the knowledge of whose logical properties is important from the point of view of the analysis of statements concerning connections; the predicates of tendencies. To explain the matter, let us give a few examples.

Assume that we observe a random motion of a large number of particles a in a certain spatial domain b. We notice that now and then some particles a form certain regular (ordered) clusters. These clusters are short-lived and do not include all particles, so that it is impossible to say that an ordering of particles a occurs in the domain b. We can say something 'weaker', namely, that there is a certain tendency towards ordering.

Another example. We expect (from experience and from all the data) a temperature increase in the domain a. And it does indeed increase, but not so rapidly and not to such high values as we have expected. It is assumed that a certain circumstance represents an obstacle to the temperature increase or even promotes its decrease. In this case it is said that there is a tendency towards a temperature decrease in a (although in reality temperature increases).

There are a variety of cases and conditions of use of such predicates of tendencies. Actually, it is possible that measurement of the degrees of tendencies occurs; in particular, there are notions of strong, moderate, weak, well-expressed, hardly noticeable, etc. tendencies. We take the existence of such predicates as a fact and consider some of their properties.

Let us introduce a special term-forming operator of tendency t. Adopt the rule: if P is a predicate, then tP is a predicate. We shall also adopt an assertion:

$$\vdash (\mathrm{P}a) \rightarrow t\mathrm{P}(a).$$

It follows from it that

$$\vdash \sim t\mathrm{P}(a) \rightarrow \sim \mathrm{P}(a).$$

For any predicate P the law of contradiction holds:

$$\vdash \sim (\mathrm{P}(a) \wedge \sim \mathrm{P}(a)),$$

and the assertion $\mathrm{P}(a) \wedge \sim \mathrm{P}(a)$ cannot be true (is logically not-true, contradictory). But the assertion

$$\vdash \sim (t\mathrm{P}(a) \wedge \sim \mathrm{P}(a))$$

is not a rule of logic, and the assertion $t\mathrm{P}(a) \wedge \sim \mathrm{P}(a)$ is logically realizable (can be true). For example, the conjunction of statements 'It is not true that the temperature of a increases' and 'The temperature of a has a tendency towards the increase' may be true.

It is also interesting to note that the attributes P and Q may exclude each other, in other words, it may be true that

$$\mathrm{P}(a) \rightarrow \sim \mathrm{Q}(a),$$

while the attributes tP and tQ may not. If P and Q exclude each other, then $\mathrm{P}(a) \wedge \mathrm{Q}(a)$ is a contradiction. But $t\mathrm{P}(a) \wedge t\mathrm{Q}(a)$ may be true. Thus, the same object can at the same time exhibit mutually excluding (in particular, opposite) tendencies. For instance, the dimensions of a certain object a can have a tendency towards increase and decrease at the same time. And there is no logical contradiction whatsoever in this. A body a can have a tendency to move in the direction b simultaneously with a tendency to move in a different direction c. The statements 'a moves in the direction b' and 'a moves in the direction c' may both turn out to be not-true, because the body a moves in a third direction d. Thus, situations are possible when

$$\sim P^1(a) \wedge \sim P^2(a) \wedge P^3(a) \wedge tP^1(a) \wedge tP^2(a)$$

is true, as well as other more complex situations which provide rich material for various verbal manipulations.

Consider a statement x such that

$$x \rightarrow (tP(a) \rightarrow Q(a));$$

then $\downarrow x$ is a condition of a realization of a tendency P, and Q is a realization of tP. Q does not always coincide with P. And it is even possible that they are incompatible, i.e., $\sim (P(a) \wedge Q(a))$.

Assume that A is a statement to the effect that there is (exists) a tendency T of a certain object. From comparison of states of this object at a certain time with those at a different time one can obtain statements B and C to the effect that the manifestation of the tendency T becomes stronger or, respectively, weaker. If $X \rightarrow A$, then $\downarrow X$ is called a state, generating T. If $Y \rightarrow B$, then $\downarrow Y$ is called a state, promoting the manifestation of T. If $Z \rightarrow C$, then $\downarrow Z$ is called a state, preventing the manifestation of T. The actual state of an object results from the combined action of states generating tendencies, and also of those promoting and preventing the manifestation of these tendencies. And in order to avoid misunderstanding, one needs in such cases to have a diverse and clearly differentiated terminology which is necessary to take into account all possible versions of and differences between, linguistic situations.

33. Paradoxes of Connections

There are such statements as

(1) $x \rightarrow (Ra)z$,
(2) $y \rightarrow (Ra)v$,

which seem to possess the following property. Applying the rules of logic, one can obtain from them:

(3) $x \wedge y \rightarrow (Ra)(z \wedge v)$,

but then the statement $x \wedge y$ may be true while $z \wedge v$ is not true, in other words, $\sim (z \wedge v)$ is true. For instance, (1) represents an expression 'If a force B is applied to a body A, this body will move in a direction C over a distance α', and (2) is an expression 'If a force D is applied to a body A, this body will move in a direction E over a distance β'; the force B and the force D can be applied simultaneously to the body A, but it cannot move simultaneously in the direction C and in the direction E (for instance, to the right and to the left). This situation is perceived as paradoxical (one of the variants of paradoxes of connections).

However, all paradoxes in this case will disappear if we reconstruct it by a sufficiently complete logical interpretation. In reality, formulations (1), (2) and (3) do not contain any indication of the conditions under which they may be accepted as true. Besides, these conditions are different. Assume that w^1 represents the conditions for (1), w^2 conditions for (2), w^3 conditions for (3). Conditions w^1 may include $\sim y$ or such z from which $\sim y$ follows, i.e., $w^1 \rightarrow \sim y$. Conditions w^2 may imply that $\sim x$, i.e., $w^2 \rightarrow \sim x$. For example, according to w^1 it may be that the force D is not applied to A, and according to w^2 it may be that the force B is not applied to A. Consider the simplest case: w^1 is $w^3 \wedge \sim y$; w^2 is $w^3 \wedge \sim x$. In order to carry out the argument, we have to have w^3 in all three cases. Then we obtain

(1) $x \wedge \sim y \rightarrow (\mathrm{R}a)z$,
(2) $\sim x \wedge y \rightarrow (\mathrm{R}a)v$,
(3) $x \wedge \sim y \wedge \sim x \wedge y \rightarrow (\mathrm{R}a)(z \wedge v)$,

(under the condition w^3 for all three of them). But since $x \wedge \sim y \wedge \sim x \wedge y$ is a contradiction, formula (3) is no longer a paradox. Now, in order to determine what consequence results from $x \wedge y$ (in particular, what will be the location of the body A if the forces B and D are applied to it simultaneously), we must either perform additional empirical research resulting in

(4) $x \wedge y \rightarrow (\mathrm{R}a)w$

or introduce a special rule for manipulating z and v by means of which it would be possible to deduce w (e.g., the parallelogram summation rule for forces).

The use of the predicates of tendencies resolves the paradoxicality of the above situation. For instance, the expression 'If a force B is applied to a body A, it will move in a direction C over a distance α (under the condition that no other forces act on A)' is replaced by a briefer one 'If a force B is applied to a body A, it will have a tendency to move in a direction C over a distance α'. In this case, no matter what forces act on

A and where it moves, our statement will record, instead of the real situation, the role of force B in it. The above statements will take the form:

(1) $x \rightarrow (\mathrm{R}a) z^*$,
(2) $y \rightarrow (\mathrm{R}a) v^*$,

where expressions z^* and v^* refer not to real situations but to tendencies. Under these conditions we obtain

(3) $x \wedge y \rightarrow (\mathrm{R}a)(z^* \wedge v^*)$,

since $z^* \wedge v^*$ is not a contradiction. The presence of opposite tendencies is not contradictory from the logical point of view. The question of how tentencies of z^* and v^* are realized should be answered, as before, on the basis of experience or of the rule specifically formulated with experience in mind.

Another version of the paradox of connections results from considering the action of one object on the other through various mediators (along various 'channels') when the result of interaction through one of the mediators excludes the result of interaction through another one.

34. Conditional Predicates

Some predicates possess properties that can be easily explained if the predicates are regarded as introduced by the following definitions:

(1) $\mathrm{P}(a) \equiv \mathrm{Df.}\ (x \rightarrow (\mathrm{R}x)\mathrm{Q}(a))$,

where x is a certain statement coordinated with $\mathrm{Q}(a)$ in such a way that

(2) $\vdash \mathrm{Q}(a) \rightarrow x$.

For instance, the predicate 'soluble in water' can be presented as a predicate according to the following scheme: '*a* is soluble in water' $\equiv$ Df. 'If *a* is placed in a container with water, *a* dissolves in water'. In this case the statement '*a* dissolves in water' is true under the condition that *a* is placed in a container with water. Thus, if the above statement is true, the expression '*a* is placed in a container with water' is also true.

Such predicates P are usually introduced as verbal modifications of Q, and the insignificant external differences among them provide a basis for confusion of terms. We shall introduce a special conditionality operator *c* for such predicates, with the following rule for construction of terms with it: if Q is a predicate, *c*Q is a predicate too.

The predicates under consideration have properties similar to those of the predicates of tendencies. Thus, the statements $x \to (\mathrm{R}x)\mathrm{Q}(a)$ and $\sim x$ are compatible, so that

(3) $(x \to (\mathrm{R}x)\mathrm{Q}(a)) \wedge \sim x$

may be true. Substituting $c\mathrm{Q}(a)$ in place of the first part of (3) according to the definition (1), we obtain:

(4) $c\mathrm{Q}(a) \wedge \sim x.$

But by virtue of (2) we have

(5) $\vdash \sim x \to \sim \mathrm{Q}(a).$

As follows from (4) and (5),

(6) $c\mathrm{Q}(a) \wedge \sim \mathrm{Q}(a).$

Thus, if (3) is true, (6) is also true.

But if the predicates Q^1 and Q^2 are mutually excluding, the predicates $c\mathrm{Q}^1$ and $c\mathrm{Q}^2$ are incompatible too. If

$$\mathrm{Q}^1(a) \to \sim \mathrm{Q}^2(a),$$

we arrive at:

$$(x \to (\mathrm{R}x)\mathrm{Q}^1(a)) \to (x \to (\mathrm{R}x) \sim (\mathrm{Q}^2(a)),$$
$$c\mathrm{Q}^1(a) \to \sim c\mathrm{Q}^2(a).$$

35. Influence

Connections, in their turn, are distinguished according to logical types. Consider some of them (from a certain point of view, the most important):

Assume that α is a formula

$$\mathrm{P}^1(a) \Rightarrow \mathrm{P}^2(a),$$

and β is

$$\mathrm{Q}^1(b) \Rightarrow \mathrm{Q}^2(b).$$

We shall say that *a* influences or exercises an influence over *b* (and *b* experiences an influence of *a*) if and only if

$$\alpha \to (\mathrm{R}\alpha)\beta,$$

where R is the temporal relation 'after this', and the time interval between

$\downarrow\alpha$ and $\downarrow\beta$ is finite. This relation is transitive. To put it differently, the temporal relation between $\downarrow\alpha$ and $\downarrow\beta$ is such that

$$\downarrow\beta > \mathrm{t}\downarrow\alpha,$$

where t is the method of establishing the temporal order, and > reads as 'later' or 'after this'. We shall also say that β is a result of the influence of α.

Let us introduce γ^i as

$$\mathrm{T}^{i1}(c^i) \Rightarrow \mathrm{T}^{i2}(c^i),$$

where c^i is an individual variable ($i = 1, 2, \ldots$). The influence of a on b will be called direct with respect to a certain class A if and only if

$$(\neg\exists c^i)((c^i \in \mathrm{A}) \wedge (\alpha \rightarrow (\mathrm{R}\alpha)\gamma^i) \wedge (\gamma^i \rightarrow (\mathrm{R}\gamma^i)\beta))$$

and mediated if and only if

$$(\exists c^1)\ldots(\exists c^n)((c^1 \in \mathrm{A}) \wedge \ldots \wedge (c^n \in \mathrm{A}) \wedge$$
$$\wedge (\alpha \rightarrow (\mathrm{R}\alpha)\gamma^1) \wedge \ldots \wedge (\gamma^n \rightarrow (\mathrm{R}\gamma^n)\beta))).$$

Individuals $c^1, \ldots, c^n$ represent mediators of influence of a on b. When $n \geqslant 1$, the expressions $\gamma^1, \ldots, \gamma^n$ will be called the mechanism of influence of a on b.

There are cases in which the influence of a on b is direct and obvious. In other cases it is realized through mediators and is determined through specification of the latter. Some general logical consequences can be derived from the above for such cases.

It is obvious that the events $\downarrow\alpha, \downarrow\gamma^1, \ldots, \downarrow\gamma^n, \downarrow\beta$ form a series, ordered in time. If this series is continuous (i.e., all its adjacent events are in contact in time), then the number of $\downarrow\gamma^1, \ldots, \downarrow\gamma^n$ is finite. Indeed, the temporal interval between $\downarrow\alpha$ and $\downarrow\beta$ (denoted by t) is finite, the duration of all events $\downarrow\gamma^1, \ldots, \downarrow\gamma^n$ and the intervals between $\downarrow\alpha, \downarrow\gamma^1, \ldots, \downarrow\gamma^n, \downarrow\beta$ is not less than a certain minimum, so that only a finite number of events can be arranged within t.

And only in the case of abstract events for which one can assume infinitesimal duration and infinitesimal intervals between events is our assertion wrong: for them, if the series of events between $\downarrow\alpha$ and $\downarrow\beta$ is continuous, then the number of events mediating them and forming the mechanism of influence is infinite.

Thus, if $\alpha \rightarrow (\mathrm{R}\alpha)\beta$ and we want to reconstruct completely the mechanism of influence of a on b, we must find a set of $\downarrow\gamma^1, \ldots, \downarrow\gamma^n$ such as is described above, this set being finite and the series $\downarrow\alpha, \downarrow\gamma^1, \ldots, \downarrow\gamma^n, \downarrow\beta$ continuous in time. Therefore, this process is not infinite: it terminates, which corresponds to cognitive practice.

But the influence of *a* on *b* may be regarded as a chain of events in space. The source of the illusion to the effect that considered from this point of view the process of determining the mediators appears infinite is the uncertainty of terms and the problem.

No matter how the spatial domains in which the events occur are arranged, the latter should be chosen in one or another way as events forming a temporal series between $\downarrow\alpha$ and $\downarrow\beta$. And then the following circumstance should be taken into account. The sum of durations of events between $\downarrow\alpha$ and $\downarrow\beta$ and the intervals between them should not exceed the duration of the interval *t*, and this means that the number of mediating events is finite (as above). If, on the other hand, this sum is greater than *t*, we have chosen a series of events such that *a* in it does not influence *b*, for example, a series missing the target.

We would like to emphasize the fact that the influence of *a* on *b* is not regarded as a motion of a certain empirical body *c*, and that the existence of the maximal velocity of motion has nothing to do with our situation (and it is exactly this existence which is referred to whenever one speaks of the existence of events not related empirically, in particular, in a causal way) to each other. The influence of *a* on *b* through motion of *c* represents a special case of influence. And for this case the following argument is indeed valid: if the variation of *b* results from the fact that *c*, moving from *a*, reaches *b*, and only from this fact, the time between $\downarrow\alpha$ and $\downarrow\beta$ should not be less than that which is required for *c* to cover the distance between *a* and *b*.

The velocity of propagation of *a*'s influence on *b* in space equals the ratio of the distance between the spatial domains in which $\downarrow\alpha$ and $\downarrow\beta$ occur to *t*. And it is impossible to obtain in this situation the same logical effect as for the velocity of motion under the assumptions made.

The question as to what extent it is possible to consider the problems concerning the effect of influence of a given event *s* on the surrounding objects, of the propagation of this influence, its decay, etc. purely from a logical point of view will not be discussed here. But it appears that something can be done with a more thorough analysis than the one conducted in this book. At any rate, it is a very intriguing issue.

Further, consider the notion of interaction having in mind that *a* influences b and b influences a. However, such formulation does not expose certain important logical details.

If $\downarrow\alpha$ and $\downarrow\beta$ represent individual events, then

$$(\alpha \rightarrow (R\alpha)\beta) \vdash \sim(\beta \rightarrow (R\beta)\alpha),$$

because $\downarrow\alpha$ is unique (it is impossible to influence the past). If b, conversely, influences a, we have

$$\beta \rightarrow (R\beta)\alpha^*,$$

where α^* is

$$P^3(a) \Rightarrow P^4(a),$$

with α^* occurring after $\downarrow\beta$, and, therefore, after $\downarrow\alpha$ (due to the transitivity of R). The reason for the illusion of simultaneous influence of a over b and of b over a is the usage of common terms of events. For instance, consider a statement 'Increase of a leads to increase of b, while increase of b, in its turn, leads to increase of a at the same time'. In this assertion the expression 'Increase of a' represents a term such that both α and α^* may be interpreted as its special cases; the expression 'Increase of b' may mean β as a result of influence of a over b, as well as a certain β^* which represents a variation of b resulting from a different source.

We have considered above a simple case of an object a exercising an influence on an object b. The complex case may be regarded as a combination of simple ones constructed according to the following scheme: assume that $\alpha^1 \Rightarrow \alpha^2, \alpha^2 \Rightarrow \alpha^3, \ldots, \alpha^n \Rightarrow \alpha^{n+1}, \ldots$ is a successive change of states of a, and $\beta^1 \Rightarrow \beta^2, \beta^2 \Rightarrow \beta^3, \ldots, \beta^n \Rightarrow \beta^{n+1}, \ldots$ is a successive change of states of b, the second sequence representing the result of the influence of the first; each state β^i corresponds to a state α^i. By definition, the state β^i occurs after the state α^i. The time interval between them is no less than the minimal. Thus, the following principles of 'retardation' hold (they are, however, purely logical): (1) it is not immediately that the result of influence of a over b can be detected; (2) if the variations of a discontinue, the variation of b proceeds for some length of time after this; this variation represents a result of the influence of a.

The explication of the terms of forces, masses, and so on, can be performed on the basis of the terms introduced. Suppose that a and b are individual variables, Q is the predicate of a certain variation or a certain tendency of variation, P is the predicate chosen specifically for introduction of the above expressions. The terms of forces can be introduced as abbreviations for such expressions as $a\downarrow(\exists b)(x \wedge y \wedge \mathrm{P}(b))$, where x is a statement 'a influences b', and y is a statement '$\downarrow \mathrm{Q}(b)$ is a consequence of $\downarrow x$'. The type of force is determined by the form of P and Q (for instance, P may represent a body's weight while Q is its displacement over a certain distance). The expression 'A force α is applied to b' in this case will be an abbreviation for $(\exists a)(x \wedge y \wedge \mathrm{P}(b))$. Thus, force represents an object which acts on another object in a certain way, the latter possessing a certain specified attribute. Explication of the mass terms proceeds according to the following scheme: the expression 'a has a mass α' is an abbreviation for the expression 'If a force β is applied to a, then $\mathrm{P}(a)$; if, on the other hand, the force β is not applied to a (including the case when the magnitude of the force is smaller than β), then $\neg \mathrm{P}(a)$'. The type of mass as well as its magnitude depends on the form of P.

The influence of a on b always lasts a finite time. It can last for the minimal time (the minimal influence) or represent a series of such minimal influences, continuous in time. The influence always results in a variation of some individuals belonging to b as a cluster – displacement in space, decay, creation (of particles), etc. Considered from this point of view, the influences may be distinguished according to the value of the velocity of the respective processes (for instance, the motion at the minimal or the maximal velocity). All this can be regarded as possible directions of development of the corresponding terminology.

In the case of influence, the continuity and discontinuity predicates are defined in a manner different from that for the logical types of objects, considered earlier:

(1) the influence of an n-tuple of individuals $(a^1, \ldots, a^n)$ $(n \geqslant 1)$ on b is continuous in t if and only if for any time interval t^i belonging to t (constituting a part of t) it can be said that at least one of $a^1, \ldots, a^n$ influences b in t^i or that t^i is minimal;

(2) the influence of an n-tuple of individuals $(a^1, \ldots, a^n)$ $(n \geqslant 1)$ on b is discontinuous in t if and only if for a certain t^i belonging to t it can be said that no one of $a^1, \ldots, a^n$ influences b in t^i, and t^i is greater than the minimal.

36. Cause

There may be many different meanings to the expression 'a is a cause of b', for instance, such as

(1) 'a is a cause of b' $\dashv\vdash$ 'If not-a, then not-b; a takes place; after this occurs b'.

(2) 'a is a cause of b' $\dashv\vdash$ 'If a occurs, after it takes place b'.

(3) 'a is a cause of b' $\dashv\vdash$ 'If there did not occur a, b would not occur; b takes place'

(4) 'a is a cause of b' $\dashv\vdash$ 'a and then b; without a there would be no b'.

In other cases, the term 'cause' implies the answer to the questions of the form 'What is the cause of an event (state, etc.) a?' And this answer may be different, in particular, as follows: (1) the cause of an event (phenomenon, state, etc.) a is everything which generates a (which leads to the emergence of a); (2) the cause of a is everything without which a cannot exist; (3) the cause of a is constituted by those special conditions that are characteristic of the appearance of a as different from the appearance of other states, and so on.

In what follows we shall formulate several theses concerning the problem of causality or, to put it more accurately, the set of problems dealing in one or another way with the word 'cause'.

It would be vain to look for some 'uniquely true' understanding of the term 'cause'. There is simply no one such meaning in existence yet. What we have are just various uses of the word, and only that. There are different epistemological situations, the fixation of which needs an exact terminology that takes into account the above differences. Thus, we can come across a relation of states such that

$$x \rightarrow (\mathrm{R}x)y$$

and such that

$$x \wedge (\mathrm{R}x)y \wedge (\sim x \rightarrow (\mathrm{R} \sim x) \sim y),$$

where R means 'then', 'after', etc. The first formula may be abbreviated to take the form

$$\downarrow x C^1 \downarrow y,$$

and the second one the form

$$\downarrow x C^2 \downarrow y.$$

Here C^1 and C^2 are special two-place predicates. And neither of them is better or worse than the other as a means of explication of the term 'cause'. And if it is absolutely necessary to use this word here, certain limitations can be introduced, such as 'positive cause', 'negative cause', 'the total cause', etc.

Furthermore, even if there were an agreement to use the word 'cause' in just one strictly defined manner, this still does not in any way guarantee that investigators will find the same causes when trying to establish the causes for the same phenomena. Thus, if we have adopted the sign C^2 as the explication of the word 'cause', this does not exclude the possibility of constructing statements $\downarrow x C^2 \downarrow y$ and $\downarrow z C^2 \downarrow y$ when determining the cause of $\downarrow y$. There is no contradiction between the above two standpoints, and according to the definition, the expression $\downarrow x$ as a cause of $\downarrow y$ is neither better nor worse than the expression $\downarrow z$, and vice versa.

Certainly, we can agree to interpret as the cause of a state $\downarrow y$ just such a state which constitutes the only cause of $\downarrow y$ (in other words, no other state, different from it, can be regarded as the cause of $\downarrow y$). But this is just general talk, without any obligations. It is one thing to define the word 'cause', and quite another to determine the causes of concrete states. When a unanimous agreement is reached in the practice of gaining knowledge to consider a certain phenomenon *b* as the cause of a certain phenomenon *a*, this is done not by virtue of the definition of the word 'cause' (no other such definition can contain an indication of the uniqueness of the cause, since there can be no guarantees of this uniqueness), but in the form of an implicit agreement to consider precisely *b* as the cause of *a*, since the relation between *a* and *b* satisfies the definition of the word 'cause' and also fulfills some other requirements that do not constitute a part of this definition (e.g., under some self-evident conditions the occurrence of *b* always leads to *a*, while non-occurrence of *b* does not lead to *a* under the same conditions; other events do not result in *a*).

The word 'cause' in the above was spoken of as a part of the predicate '...cause...'. But it is used as a subject as well or, to put it more precisely, as a part of subjects of the form 'the cause of an event', 'the cause of $\downarrow x$', 'the cause of the fact that *x*', etc. It is defined in this role as a derivative of the predicate 'cause' in the following manner (the letters *x* and *y* represent variables for state terms): 'The cause of *x*' ≡ Df. 'A state $\downarrow y$ such that $\downarrow y$ is the cause of $\downarrow x$' (in other words, the cause of a certain state represents another state such that it is its cause). The agreement described in the preceding section represents an implicit definition of this kind, containing in addition an element of limitation.

The well-known inductive methods of establishing the causal relations (the Bacon-Mill methods) cannot be regarded simply as research techniques based on an assumption that the term 'cause' is defined before they are applied and independently of it. They themselves represent an implicit definition of various cases of usage of the term 'cause'. Thus, the method of concomitant variations is formulated as follows: if every time when an event *a* occurs, an event *b* occurs after this, then *a* is the cause of *b*. This should be understood not as an assertion containing the term 'cause' the meaning of the latter being known even without this assertion, but as a fragment of an implicit definition of the very word 'cause'. And the more precise form of this method should be this: if every time after the occurrence of *a*, *b* does occur, then *a* will be called the cause of *b*. In a different form, '*a* is the cause of *b*' $\dashv\vdash$ 'If *a* occurs, then *b* does occur'. The formulation of the method of exclusive similarity is as follows: if the cases for which *b* occurs differently in respect to everything, the only similarity between them being that the occurrence of *b* precedes that of *a*, then *a* is the cause of *b*. Again, this is not just a matter of research techniques, but a fragment of the implicit definition of the word 'cause'. The argument for other methods is similar. If, on the other hand, the above methods are considered to be techniques for studying causes, with the meaning of the word 'cause' known without them, the confusion is inevitable. It is this confusion that has led to frequent criticism of the above methods in the history of logic and philosophy. In such situations they were faced with excessive claims which were unjustified.

Furthermore, when formulating the explication of terms, one has to distinguish between wishful thinking and the actual capabilities of a language. Thus, we can formulate the problem of defining the cause of *a* in such a way that everything which leads to *a*, and without which the emergence of *a* is impossible, will be called the cause of *a*. But this is only our desire and not more than that (similar to the recommendation of a deep and diversified study of *a*). Then questions arise of the following sort: what shall we do with the World, without which there can be no state of affairs whatsoever?; what to do with the Galaxy, without which no events in the Solar system can occur?: what to do with electrons, when we discuss the causes of defeat in this or that war? and so on. No scholastic trickery will be of any use here. From the point of view of practice, an explication which is adequate to the wish mentioned is simply impossible.

If *a* is the cause of *b*, the occurrence of *a* precedes that of *b* in time. The temporal relation between *a* and *b* is one of the attributes of the causal relation, which participate in the definition of the latter. While the whole terminology of time is defined and introduced into usage independently of the term 'cause', the reverse is not true.

Assume that the following agreement has been reached: the expression '*a* is the cause of *b* in coordinates *c*' has a meaning which is identical to that of the expression 'Always, in coordinates *c*, *b* occurs after the occurrence of *a*', where *a* and *b* represent variables for event terms (or represent terms of arbitrary events), whereas *c* is a variable for coordinate terms (or arbitrary coordinate terms).

If a person N is in favor of this agreement, she decides to apply the term 'cause' with a meaning that has been established by this definition. But suppose that, observing a certain concrete set of objects, N has arrived at the conclusion: 'A is the cause of B in coordinates C', where A, B, C are quite specific terms. If we know the meaning of the word 'cause', we are not guaranteed against the possibility that this conclusion might not be true. If it represents a result of observation of a great number of cases corresponding to the sequence of events A and B in time in coordinates C, it is not impossible that at some time the following case will be discovered: after the occurrence of A in coordinates C, B does not occur. This case disproves the above conclusion of N without rejecting the definition of the word 'cause' nor does it disprove the fact that it is known to N.

As a rule, these two aspects – the terminological and the heuristic – are confused. The description and interpretation of Bacon–Mill methods represent a classical example of such confusion.

Now imagine the following situation: (1) the word 'cause' is used with most diverse meanings; (2) these meanings are vague and unstable; (3) the terminological and the heuristic aspects of its application are confused. Thus, nearly every statement including this word is either totally meaningless or has a very vague meaning corresponding to the concrete nature of the context or the situation in which it is used.

In this connection, a remark is in order that in general the process of research (making discoveries) does not coincide with the operations of establishing the meaning of terminology, if these operations are governed by the rules of logic and conducted at the level of logic. In the course of research, terminology may be produced and is produced; it remains for some time (not infrequently, always) closely associated with the research. Logical explication of this terminology should first of all separate it from this genetic source, and consider terminology independently of it. And it is all the more important to find the logical differences here, the stronger and the finer this association is.

The situation with the term 'cause' is quite characteristic and instructive. In its amorphous state, the term 'cause' has been and is quite useful both in its own time and now (just as the term 'schizophrenia' in medicine, denoting the most diverse forms of illness, is useful as a means for preliminary orientation). But if the linguistic fragments containing this term are studied in greater detail, it will be discovered that the description of these fragments in the framework of logic requires a more precise and differentiated terminology which either renders this term generally unnecessary or attaches to it only one of its multiple meanings. We would be entirely justified to apply the above considerations to the whole set of general terms in the methodology of science.

37. Forms of Causal Relations

Comparison of those cases when the expression '$\downarrow x$ is the cause of $\downarrow y$' is used reveals the following: sometimes it is assumed that y does not follow logically from x; in other cases no such assumption is made; sometimes it is supposed that from '$\downarrow x$ is the cause of $\downarrow y$' and x, follows y; in other cases this is not done; sometimes it is assumed that from '$\downarrow x$ is the cause of $\downarrow y$' and $\sim x$, follows $\sim y$, and sometimes not; sometimes one assumes the transitivity of the causal relation, sometimes not, etc. Therefore, it would be completely hopeless to look for a certain 'invariant' in these cases of usage, which could be interpreted as the 'true understanding of cause'.

It seems appropriate to us to speak of the forms of causal relations between states $\downarrow x$ and $\downarrow y$ including in their number the following expressions (where in all formulas $R^1 a$ reads as 'after a', and $R^2 a$ as 'before a', 'preceding a'):

$$x \rightarrow (R^1 x) y,$$
$$y \rightarrow (R^2 y) x,$$
$$\sim x \rightarrow (R^1 \sim x) \sim y,$$
$$\sim y \rightarrow (R^2 \sim y) \sim x,$$
$$(x \rightarrow (R^1 x) y) \wedge y \wedge (R^2 y) x,$$
$$(y \rightarrow (R^2 y) x) \wedge x \wedge (R^1 x) y,$$
$$(\sim x \rightarrow (R^1 \sim x) \sim y) \wedge x \wedge (R^1 x) y,$$
$$(\sim y \rightarrow (R^2 \sim y) \sim x) \wedge x \wedge (R^1 x) y,$$
$$(x \rightarrow (R^1 x) y) \wedge (y \rightarrow (R^2 y) x),$$

etc. The common element in all these statements is the presence of a conditional relation between statements describing states, and a temporal relation 'after this' or 'before this'. And the properties of the corresponding statements completely determine those of these relations (the former are considered in the second chapter). In particular,

$$(x \rightarrow (R^1 x)y) \vdash (\sim y \rightarrow (R^2 \sim y) \sim x),$$
$$(x \rightarrow (R^1 x)y) \wedge (y \rightarrow (R^1 y)z) \vdash (x \rightarrow (R^1 x)z).$$

If *a* is the cause of *b*, then *b* is a consequence of *a* (this is a definition of the term 'consequence').

If $\downarrow x$ and $\downarrow y$ are individual terms, then the term 'cause' is defined for them in the following way: $\downarrow x$ is the cause of *y*' $\dashv\vdash (\downarrow x \rightharpoonup a) \wedge \wedge (\downarrow y \rightharpoonup b) \wedge$ '*a* is the cause of *b*' (i.e., $\downarrow x$ is the cause of $\downarrow y$ if and only if $\downarrow x$ is a, $\downarrow y$ is *b* and *a* is the cause of *b*). Thus, it is implied here that the term 'cause' is first defined for general terms, and then for individual ones. And this is sensible since the causal relation for general terms is defined through the conditional relation of statements, while the latter implies generalizations, in particular, such as

$$(x \rightarrow (Rx)y) \dashv\vdash (\forall \downarrow x)(Rx)y,$$

etc. And even from the point of view of establishing the causal relations between individual states $\downarrow x$ and $\downarrow y$ an additional element is required besides the fact that the second state appears in time after the first; this additional assertion should be valid for classes of states the elements of which are $\downarrow x$ and $\downarrow y$.

Incidentally, this explains why the majority of statements concerning the causal relation between past states can neither be confirmed nor disproved, since it is impossible to construct true statements such as $x \rightarrow (Rx)y$ for them. In this case, again, instead of using a logical argument to convince the reader, implicit agreements are made to consider some events the cause of others, in other words, statements of the form '*a* is the cause of *b*' are implicitly adopted as postulates with various kinds of talk accompanying this decision. This talk is then called substantiation of the decision made.

Certain relations, including in one way or another the probability of events, represent special cases of causal relations between $\downarrow x$ and $\downarrow y$. Those are, e.g., the relations fixed through such statements as 'The probability of the fact that $\downarrow x$ is the cause of $\downarrow y$ is α (is equal to α, is greater than α, etc.)', 'If *x*, then after it $\downarrow y$ occurs with a probability α', etc.

We shall call statements of the form '*x*, and therefore (for this reason, because of this) *y*' and '*y* for the reason that (because) *x*' in which *x* and *y* represent expressions concerning individual empirical states or events, causal conjunctions, and we shall write them briefly in symbolic form as

$$x \rhd y.$$

There are examples of such statements: 'The ship "Titanic" sank because it collided with an iceberg', 'The presence of the enemy's agent in the headquarters of the *N*th division is the cause of the fact that the enemy is in possession of secret information about the division'.

Causal conjunctions $x \rhd y$ have the following structure: (1) statement x fixes what happened (or is happening) in terms of an empirical event, or that a certain empirical state took (or is taking) place; in particular, the latter may represent a cluster of empirical events (sometimes, uniform) in space or in time; (2) statement y fixes another empirical event or state; (3) there is yet another causal ingredient on the basis of which we can say that the first event (state) is the cause of the other. This causal ingredient may have different logical forms. It can be expressed by the statement 'If an event occurs (a state takes place) from a class a which includes an event (state) $\downarrow x$, after this an event occurs (a state takes place, happens, etc.) from a class b which includes $\downarrow y$'. But this is not always so. Sometimes the causal ingredient is fixed by a statement 'If $\downarrow x$ had not occurred, $\downarrow y$ would not occur', in other cases by a statement 'All other possible causes of $\downarrow y$, except $\downarrow x$, should be rejected', etc. Sometimes the causal ingredient may be regarded as a result of habit or is introduced through observation of a sequence of events generating one another and is not expressed in language in any form. To put it briefly, it is in principle impossible to find a universal causal ingredient for z such that $x \rhd y$ can be presented as an abbreviation for $x \wedge y \wedge z$.

Internal negation of causal conjunctions $x \urcorner \rhd y$ means that while $x \wedge y$ is accepted, the presence of the causal ingredient is rejected. External negation of $x \rhd y$ is equivalent to $\sim x \vee \sim y \vee (x \urcorner \rhd y)$.

We shall adopt the following logical rules for causal conjunctions, providing an implicit definition of the expression $\rhd$ ('and therefore', 'for this reason', and so on):

(1) $\vdash (x \urcorner \rhd x)$,
(2) $(x \rhd y) \vdash x \wedge y$,
(3) $(x \urcorner \rhd y) \vdash x \wedge y$,
(4) $(x \rhd y) \vdash (y \urcorner \rhd x)$,
(5) $\sim (x \rhd y) \dashv\vdash \sim x \vee \sim y \vee (x \urcorner \rhd y)$,
(6) $(x \rhd y) \wedge (y \rhd z) \vdash (x \rhd z)$,

(7) $(x \triangleright y) \wedge (x \triangleright z) \vdash (x \triangleright y \wedge z)$,
(8) $(x \triangleright y) \wedge (z \triangleright v) \wedge ((y \wedge v) \triangleright w) \vdash ((x \wedge z) \triangleright w)$,
(9) $(x \triangleright y) \wedge (y \rightarrow z) \vdash (x \triangleright z)$,
(10) $(x \rightarrow y) \wedge (y \triangleright z) \vdash (x \triangleright z)$,
(11) $x \wedge y \wedge (\downarrow x \rightharpoondown \downarrow a) \wedge (\downarrow y \rightharpoondown \downarrow b) \wedge (a \rightarrow (\mathrm{R}a)b) \vdash (x \triangleright y)$,

where R is 'after this'.

38. Determinism and Indeterminism

The concept of determinism (and its negation, indeterminism) is ambiguous. There are at least the following different meanings to it:

Determinism is understood first of all as a principle to the effect that for every empirical state (event) there is another state which represents its cause (everything that happens has a cause, or nothing occurs without a cause). The following assumption may be regarded as a modification of this principle: for every empirical state we can find another state which represents its cause (everything has a cause). According to these assumptions, not only are there causes for any states, but there is a possibility of their determination. Obviously, it is stronger than the previous one.

In its turn, indeterminism is interpreted as the negation of the above principles, with the negation of the first meaning the assumption of the existence of states that have no causes (causeless events). On the other hand, the negation of the second means only an assumption to the effect that sometimes it is impossible to find the causes of events.

Further, determinism is understood as an assumption according to which for every empirical state $\downarrow x$ there is another state $\downarrow y$ such that $y \rightarrow (\mathrm{R}y)x$ with a probability 1; here R represents a certain temporal relation. In terms of a and b (the variables for statements concerning the empirical objects) this assumption can be written as follows:

$$(\forall a)(\exists b)(b \rightarrow (\mathrm{R}b)a).$$

A stronger assumption results if we suppose that for every empirical state $\downarrow x$ not only it is possible to find the above state $\downarrow y$ but it can be detected, too. On the other hand, the negation of determinism in this interpretation is understood as an assumption according to which there are cases when $y \rightarrow (\mathrm{R}y)x$ is impossible to be found with a probability of 1, but only with a probability less than 1. And this represents an entirely different aspect of the matter.

There is also a more narrow meaning of the word 'determinism', namely, it is understood as a principle according to which, if a given sphere of the world is characterized at a given time by sufficiently complete and accurate information, its state in the future can be predicted. Physical determinism represents a still more narrow principle: if one knows the momentum and trajectories of physical bodies at present, it is possible to predict their locations in future. We shall discuss the issue of this type of determinism again below.

Let us adopt the following definitions:

(1) We shall say that an event $\downarrow y$ is deterministic with respect to a class of events A if and only if

$$(\exists \downarrow x)((\downarrow x \in \mathrm{A}) \wedge \sim(\downarrow y \in \mathrm{A}) \wedge (x \rightarrow y)),$$

where x is a variable for statements.

(2) We shall say that $\downarrow y$ is not deterministic with respect to A if and only if

$$(\forall \downarrow x)((\downarrow x \in \mathrm{A}) \wedge \sim(\downarrow y \in \mathrm{A}) \rightarrow \sim(x \rightarrow y)).$$

As the expressions 'is inside', 'is outside' are defined, the following definitions are legitimate:

(3) Events, occurring inside an individual *a*, are internal events for *a*.

(4) Events, occurring outside an individual *a*, are external events for *a*.

If the class A appearing in definitions (1) and (2) represents a class of internal (external) events for $\downarrow y$, a definition of the expressions 'internally (externally) determined (not determined)' results (corresponding to definitions (5) and (6)).

Finally, let us adopt the following definitions:

(7) An event is free if and only if it is not determined both externally and internally.

(8) An event is not free if and only if it is not determined externally or internally.

(9) An event is internally free (not free) if and only if it is internally not determined (determined).

(10) An event is externally free (not free) if and only if it is externally not determined (determined).

A consequence of the definitions is that if an individual has minimal dimensions, it is internally free (or that an event occurring inside it is impossible).

39. Other Forms of Connections

In language there is an enormous number of expressions that denote empirical connections: 'dependence', 'effect', 'inseparably connected', 'compatibility', and so on. Most frequently they are amorphous, have multiple meanings, depend on the context and the specific features of the medium of usage, etc. Of course, a hope to establish a rigorous logical order here would be naive. We shall just draw the attention of the reader to two circumstances: (1) analysis of these expressions reveals one or another combination of such statements as $x \to (Rx)y$; (2) these combinations are quite diverse, and it is only arguments of practical suitability that determine their number and their degree of stability.

For instance, compatibility of objects may be defined in the following manner: a and b are compatible if and only if always and everywhere the assertion

$$(\mathrm{E}(a) \leftrightarrow \mathrm{E}(b)) \wedge (\neg \mathrm{E}(a) \leftrightarrow \neg \mathrm{E}(b))$$

holds. One can define the compatibility of attributes as follows: P and Q are compatible if and only if for any object a the assertion

$$(\mathrm{P}(a) \leftrightarrow (\mathrm{Q}(a)) \wedge (\neg \mathrm{P}(a) \leftrightarrow \neg \mathrm{Q}(a))$$

holds.

The expression 'dependence' ('depends') is in frequent use. It has multiple meanings. In particular, one can speak of the dependence of a state upon another state, of a state upon a non-differentiated object, of an object on a state, of an object upon an object. And each case, in its turn, allows for variations. For example, the following definitions are possible: (1) $\downarrow x$ depends on $\downarrow y$ if and only if x under condition z and $\sim x$ under condition $z \wedge y$; (2) $\downarrow x$ depends on $\downarrow y$ if and only if x under condition z and $\sim x$ under condition $z \wedge \sim y$. Thus, the determining part of (1) contains $z \wedge y$, and that of (2) $z \wedge \sim y$. In addition, the dependence is sometimes interpreted as influence, interaction, the causal relation, etc. Finally, the word 'dependence' is used with the following meaning: (1) $\downarrow x$ depends on $\downarrow y$ if and only if $(y \to x) \vee (y \to, \sim x)$; (2) a depends on b if and only if $(\exists \mathrm{P})(\exists \mathrm{Q})$ '$\downarrow \mathrm{P}(a)$ depends on $\downarrow \mathrm{Q}(b)$'; (3) combinations of cases such as (1) and (2).

The connections of dependence differ from those of transfer. Let us cite right now one of their possible explications: (1) an expression 'An object a gives (transfers) x to an object b' represents an abbreviation for the expression 'If a loses x, b acquires this x at the same time', where the expression 'a, loses x' is, in its turn, an abbreviation for the expression '$x \in a$ at t^1, and then $x \neg\in a$ at t^2', whereas 'b acquires x' represents an abbreviation for '$x \neg\in b$ at t^1, and then $x \in b$ at t^2'; (2) the exchange is defined as 'a gives x to an object b, and b gives y to an object a at the same time'; (3) the expression 'a transfers x from b to c' is an abbreviation for the expression 'b transfers x to an object a, and then a gives x to an object c'. As we see, in this situation also we are dealing with a conditional connection and an order relation.

There are genetic connections (creation, transformation, and so on) the explication of which is characterized by a more complex structure. However, in such cases we have to keep in mind that there is a possibility of expressions here that seem to belong the realm of empirical connections but lose this appearance during explication. This refers, for instance, to the expression 'a resulted from b'. Let us consider it in greater detail.

The expression 'a separated out of b' can be defined as

$$(\exists t^1)(\exists t^2)([(a \in b)t^1] \wedge \mathrm{E}t^1(a) \wedge \mathrm{E}t^1(b) \wedge [(a \neg\in b)t^2] \wedge \\ \wedge \mathrm{E}t^2(a) \wedge \mathrm{E}t^2(b)),$$

where t^1 and t^2 are the variables for time. The expression 'a separated out of b' may be defined as 'a separated out of a cluster formed of a and b'. The expression 'a resulted from b' may be defined for individual terms in the following manner:

$$(\exists x)(\mathrm{P}(x) \wedge \mathrm{Q}^1(a,x) \wedge \mathrm{Q}^2(b,x) \wedge ((y \in x) \rightarrow (\exists z)((z \in x) \wedge \\ \wedge \mathrm{R}(y,z)))),$$

where x is the variable for empirical series, $\mathrm{P}(x)$ is 'x is continuous', $\mathrm{Q}^1(a,x)$ is 'a is the end of x', $\mathrm{Q}^2(b)$ is 'b is the beginning of x', y and z are individual variables. For general terms this expression is defined as

$$(\alpha \in \mathrm{K}a) \rightarrow (\exists \beta)((\beta \in \mathrm{K}b) \wedge \mathrm{T}(\alpha, \beta)),$$

where α and β are individual variables, $\mathrm{T}(\alpha, \beta)$ is 'β resulted from β'. For predicates, the expression 'P resulted from Q' is defined as

$$(\exists \alpha)(\exists \beta)\mathrm{T}(\alpha \downarrow \mathrm{P}, \beta \downarrow \mathrm{Q}).$$

40. On the Logical Situation in Microphysics

The idea of a special logic of the microscopic world (the logic of microphysics, the logic of quantum mechanics), which is different in principle from the ordinary logic that was formed on the basis of the study of phenomena in the macroscopic world, resulted from the peculiar properties and conditions of the analysis of the microworld phenomena as compared to those in the analysis of the macroworld phenomena.

When trying to justify the thesis of a special logic for microphysics, authors refer, on the one hand, to the exclusive and unusual properties of objects of the microworld, and, on the other, to the exclusive and unusual conditions of their cognition. Logic, however, is not in general a theory of being, and its rules are the same for all sciences, regardless of the spheres of being they are concerned with. The only factor to result from a difference in the spheres of being (i.e., the areas of cognition) is that in different situations terms and statements with different logical structures may appear. Certainly, different rules of logic may be applied in such cases. But this does not mean at all that one and the same rule of logic in one case leads to positive results, and in another to errors. If a situation develops in a certain sphere of science so that it seems that the application of some logical rules leads to errors, it is desirable to find the source of confusion in a mixing of various logical operators, in the absence of their necessary differentiation or in their incorrect (out-of-place) usage rather than to yield to doubts concerning the universality of these rules or to construct a special logic for this branch of science.

It is believed that the logical situation in microphysics is unusual because it contains statements that are neither true nor false: they are characterized by a third truth value ('uncertain').

Besides, there are pairs of statements in microphysics which are interrelated in such a way that if one of the statements is true or false, the other is uncertain (complementary statements).

But the three-valuedness of statements does not constitute an exclusive specific feature of microphysics, and the ideas of many-valued logic were known to logicians before the appearance of quantum mechanics while the rules of three-valued logic are as universal as those of two-valued logic, regardless of whether they may be applied in certain cases or not.

The situation with complementary statements is similar. We can find this kind of relation between statements not only in microphysics. But even if the complementary statements had constituted an exclusive privilege of microphysics, their discovery would not mean any revolution in the methods of reasoning, just as the discovery of the three-valuedness of statements in general would not. Let us explain this in more detail.

As was already noted in the first chapter, in order to establish the rules of inference as well as those of their application, we do not need to know at all how many truth values the statements are assigned. By their very nature, these rules are such that it is sufficient for their establishment and application to manipulate only one truth value 'true' and its negation 'not-true'. And since all the predicates of the truth values can be reduced to 'true', every description based on three truth values can be replaced by a description adequate to it, which includes only the value 'true' and its negation.

The situation in microphysics that is responsible for the idea of a special logic of microphysics is as follows:

(1) in order for a statement concerning the momentum of a microparticle to be accepted or rejected, it has to be measurable; a similar statement is valid for coordinates;

(2) on the other hand, if there is a possibility of accurately measuring the momentum (coordinates) of a particle, then accurate measurement of the coordinates (the momentum) turns out to be impossible.

Thus, in the description of this situation, predicates of truth values have not been used at all, which is in complete agreement with our assertion to the effect that these predicates can be completely eliminated from language. But these predicates are used, and this is what constitutes the basis for the discussion.

Let us introduce abbreviations:

(1) P(a) – 'The momentum of a particle a can be accurately measured';

(2) Q(a) – 'The coordinates of a particle a can be accurately measured';

(3) S(a) – 'The momentum of a equals α';

(4) T(a) – 'The coordinates of a equal β';

The following statements are valid for the assertions P(a) and Q(a) and their negations $\neg$P(a) and $\neg$Q(a):

(1) x is true if and only if indeed x (where x is any of them);

(2) P(a) is false if and only if $\neg$P(a) is true; $\neg$P(a) is false if and only if P(a) is true; the statement for Q(a) and $\neg$Q(a) is similar;

(3) the negation of truth ('not-true') for such assertions coincide with falsity (they are two-valued).

The following statements are valid for assertions S(a) and T(a) and their negations $\neg$S(a) and $\neg$T(a):

(1) S(a) is true if and only if P(a) is true and indeed S(a); similarly, T(a) is true if and only if Q(a) is true and indeed T(a); $\neg$S(a) is true if and only if P(a) is true and indeed $\neg$S(a); $\neg$T(a) is true if and only if Q(a) is true and indeed $\neg$T(a); these statements differ from those of the first paragraph

for P(a) and Q(a) in that P(a) is implied as the condition of truth of S(a) and $\neg$S(a), and Q(a) is implied as the condition of truth of T(a) and $\neg$T(a).

(2) S(a) is uncertain if and only if P(a) is false; $\neg$S(a) is uncertain if and only if P(a) is false; T(a) is uncertain if and only if Q(a) is false; $\neg$T(a) is uncertain if and only if Q(a) is false; here, unlike the case of P(a) and Q(a), S(a), T(a) and their negations can be checked depending on the truth and falsity of other statements, so that there is a possibility of introducing the third value 'uncertain';

(3) S(a) is false if and only if $\neg$S(a) is true; $\neg$S(a) is false if and only if S(a) is true; T(a) is false if and only if $\neg$T(a) is true; $\neg$T(a) is false if and only if T(a) is true.

P(a) is related to Q(a) in the following manner: if one of them is true, the other is false.

From the above premises one obtains a conclusion: if one of the pair of S(a) and T(a) is true or false, the other is uncertain.

Let us adopt a definition. Two statements x and y are called complementary if and only if they are related in the following way: if one of them is true or false, the other is uncertain.

According to this definition and the assertion, obtained earlier, statements S(a) and T(a) represent complementary statements.

It is obvious that the conjunction $x \wedge y$ of two complementary statements x and y cannot be true, since by its very definition it is false if one of x and y is false; if, on the other hand, one of x and y is true, then $x \wedge y$ is uncertain due to the uncertainty of the other element of the pair x and y; it is uncertain if both x and y are uncertain.

The above represents a description of the logical situation of the statements S(a) and T(a) in terms of three-valued logic. But the same description can be achieved in terms of two-valued logic with the values 'true' and 'not-true' as well. In order to do this, it is necessary to replace expressions with terms 'uncertain' and 'false' everywhere by those adequate to them containing the terms 'true' and 'not-true'. Namely, the following should be done:

(1) according to the condition, for statements P(a), Q(a) and their negations the uncertainties are excluded, and the expressions 'x is false' are replaced by 'x is not-true'.

(2) expressions 'x is uncertain' are replaced in the case of statements S(a) and $\neg$S(a) by expressions 'P(a) is not-true' or '$\neg$P(a) is true', and for statements T(a) and $\neg$T(a) by expressions 'Q(a) is 'not-true' or '$\neg$Q(a) is not-true':

(3) the expression 'S(a) is false' is replaced by '$\urcorner$S(a) 'is true', '$\urcorner$S(a) is false' by 'S(a) is true', 'T(a) is false' by '$\urcorner$T(a) is true', '$\urcorner$T(a) is false' by 'T(a) is true'.

If these substitutions cannot be carried out, the terms 'uncertain' and 'false' remain meaningless.

Besides, we can describe the above logical situation without employing the truth value terms, and in a very simple way at that.

Let us adopt the following definition. Statements $Q^1(b)$ and $Q^2(b)$ will be regarded as complementary if and only if statements $P^1(b)$ and $P^2(b)$ are coordinated with them, such that

(1) $P^1(b) \rightarrow \sim P^2(b)$,
(2) $Q^1(b) \rightarrow P^1(b)$,
(3) $Q^2(b) \rightarrow P^2(b)$,
(4) $\urcorner Q^2(b) \rightarrow P^1(b)$,
(5) $\urcorner Q^2(b) \rightarrow P^2(b)$.

It follows from assertions (1)–(5) that

(6) $P^2(b) \rightarrow \sim P^1(b)$,
(7) $\sim P^1(b) \rightarrow \sim Q^1(b)$,
(8) $\sim P^2(b) \rightarrow \sim Q^2(b)$,
(9) $\sim P^1(b) \rightarrow \sim \urcorner Q^1(b)$,
(10) $\sim P^2(b) \rightarrow \sim \urcorner Q^2(b)$.

From (1), (2) and (4) we obtain:

$$Q^1(b) \vee \urcorner Q^1(b) \rightarrow \sim P^2(b).$$

From (6), (3) and (5) it follows that

$$Q^2(b) \vee \urcorner Q^2(b) \rightarrow \sim P^1(b).$$

From (8) and (10) one obtains:

$$\sim P^2(b) \rightarrow \sim Q^2(b) \wedge \sim \urcorner Q^2(b).$$

According to (7) and (9),

$$\sim P^1(b) \rightarrow \sim Q^1(b) \wedge \sim \urcorner Q^1(b).$$

Finally, we have:

$$Q^1(b) \vee \urcorner Q^1(b) \rightarrow \sim Q^2(b) \wedge \sim \urcorner Q^2(b),$$
$$Q^2(b) \vee \urcorner Q^2(b) \rightarrow \sim Q^1(b) \wedge \sim \urcorner Q^1(b).$$

Introducing the uncertanity operator, we obtain:

$$Q^1(b) \vee \neg Q^1(b) \rightarrow ?Q^2(b),$$
$$Q^2(b) \vee \neg Q^2(b) \rightarrow ?Q^1(b),$$

and further,

$$\sim ?Q^1(b) \rightarrow ?Q^2(b),$$
$$\sim ?Q^2(b) \rightarrow ?Q^1(b).$$

Thus, if we are dealing with complementary statements, the negation of the uncertainty of one of them leads to the uncertainty of the other.

If x and y are complementary statements, then $\vdash \sim (x \wedge y)$. Indeed, consider

(1) $$Q^1(b) \wedge Q^2(b).$$

Since by virtue of (1) and (2) we have

(2) $$\vdash Q^1(b) \wedge Q^2(b) \rightarrow P^1(b) \wedge P^2(b),$$

by making use of (6) we obtain

(3) $$\vdash Q^1(b) \wedge Q^2(b) \rightarrow P^1(b) \wedge \sim P^1(\mathrm{b}).$$

And since, as follows from D^{13} and U^3,

(4) $$\vdash \sim (P^1(b) \wedge \sim P^1(b)),$$
(5) $$\vdash \sim (P^1(b) \wedge \sim P^1(b)) \rightarrow \sim (Q^1(b) \wedge Q^2(b)),$$

by virtue of (4) and (5) we have

$$\vdash \sim (Q^1(b) \wedge Q^2(b)).$$

Further, it is known that the objects of the microworld cannot be observed by themselves. One can observe just the results of their interaction with the objects of the macroworld (their traces in the macroworld). Let us formulate this situation in a form convenient from the point of view of the logical approach to it.

Assume that objects of a class A (a universe A) are studied under the following conditions: (1) we can observe the results of action of objects of A on objects of a class B (on a universe B), i.e., observe traces of A in B; (2) traces of A in B are observed as phenomena of universe B; (3) without universe B it is impossible to observe universe A

We shall assume here that the set of terms for denoting phenomena in universe B and for constructing statements concerning them (in other words, the language for B) is given. The problem consists in formulating the language for A. A logically correct way for constructing the language for A is as follows: (1) elementary subject-terms are introduced (implicitly or explicitly) according to the scheme: 'The objects of A, having traces α in B, will be called β', where α is a description of the traces of objects of A in B as phenomena of B in the language for B; (2) the elementary predicate-terms are introduced according to the scheme: 'We shall say that y if and only if $z^1, \ldots, z^n$', where y is a statement containing the defined predicate, and $z^1, \ldots, z^n$ are statements in the language for B, fixing the traces of A in B; (3) derived terms are introduced by making use of the elementary terms according to the general rules of logic; (4) measurements of traces of A in B result, respectively, in the predicates of the quantities for the language for A.

Due to the logical properties of definitions of terms and the rules for constructing derived terms, statements of the language for A can be replaced completely by statements of the language for B, which do not contain terms of the language for A. All this can be done without loss of information about the universe A (in analogy with the absence of the loss of information when the word 'diamond' is replaced by the expression 'equilateral quadrangle').

The picture of universe A, formulated in a language of the above type, is a fully scientific picture. Moreover, if we desire to find a particular scientific picture of universe A irrespective of universe B (the universe A as such, by itself, etc.), we have to take the following circumstance into account. If the objective is purely scientific, it should be realized, presumably, by a certain theoretical construction (a theory or a mental model) which provides the possibility of deduction and predictions in a given sphere of science, or through a certain set of heuristic principles producing an orientation for experimental research.

Assume that X is such a theoretical construction, formulated in a special language for A without making use of the language for B. Leaving alone the question of how this can be done in practice, let us consider a different question: how can X be used as a means for deduction and prediction as far as the traces of A in B are concerned? In other words, by applying X to given statements Y in the language for B how can we obtain new statements Z in the same language? Only the following method is logically imaginable: (1) statements Y are translated into the language in which X is constructed according to special correspondence rules; (2) X is applied to the statements thus obtained, (Y*) to deduce the statements Z*; (3)

according to the same rules of correspondence the latter are translated into the language for B, which results in Z.

But in order for X to give correct predictions (confirmed by experiment) in the language for B, it should be constructed in such a way that it is possible to translate at least a part of it (according to the above rules of correspondence) into the language for B, and the resulting theoretical construction X* would provide a possibility of deducing Z from Y without their mediating translations from Y* into Z*. Thus, the construction of X represents just a special case of formulation and use of theoretical constructions. It is possible that this method has certain advantages. But it is not necessary in principle. And the main point is that it creates a phraseological illusion of independence of the picture of universe A in the language for B (provided, of course, that one remains in the frame of pure science and ignores other objectives – those of propaganda, education, etc.). The situation with the system of heuristic principles is similar; under the above conditions their application to universe A is unthinkable without a language for A, constructed in the above way.

41. The Wave-Particle Duality

The notion of the incompatibility of particle (corpuscular) and wave properties in one individual resulted from conventional usage of words. For such usage the particle was given spatial localization and compactness of all its parts, and the wave that of propagation of a certain process in space. As a typical example, we can mention a stone and a wave on the water propagating from the location of the stone's fall. But there is a considerable difference between the usage of words on the basis of sensory perceptions and the construction of rigorous definitions of these words. The following comes to light during an attempt to define the expressions 'wave' and 'particle' in correspondence with the rules for constructing definitions and with their intuitive meaning (conventional usage):

Definitions of the expressions 'particle' and 'wave' can be formulated in such a way that assertions of A will be logically deducible from them: (1) 'If an individual is a particle, it is not a wave'; (2) 'If an individual is a wave, it is not a particle'. If the construction of the definitions is logically correct, an assertion B logically follows from A: 'An individual which is simultaneously a particle and a wave is impossible (does not exist)'. And there can be no exceptions to B whatsoever. If, on the other hand, we accept for some individuals that they are particles and waves at the same time, we thereby reject B and reject A. Therefore, the definitions of the expressions 'particle' and 'wave' should be formulated in such a way that it would be impossible to deduce assertion A from them. Any other conclusions from a given situation are logically illegitimate. If the linguistic situations is such that A is accepted but B is rejected, and if in all other respects this situation is logically correct, the above circumstance provides clear evidence of the fact that the linguistic expressions in use are handled poorly from the logical point of view.

However, it is possible to define the expressions 'particle' and 'wave' in such a way that the assertions 'Certain individuals represent particles and waves at the same time' and 'An individual is possible which is a particle and a wave' would be logically consistent (logically realizable), while the assertion A will be unprovable. Moreover, even a weaker statement: 'An individual is possible for which A holds' will be unprovable. The very basis of this definition would correspond to the intuitive meaning of the words 'particle' and 'wave' and would imply nothing extralogical besides this intuitive meaning. And there is no element in this effect such that it can be interpreted as a result of a certain premeditation. The point is just that in the explication of linguistic expressions according to the rules of logic, we remove a covering shell from them which has been formed in connection with the ways these expressions function in a complex system of social, psychological, and other links and which have no positive meaning whatsoever if considered from a purely scientific point of view.

Let us adopt the following definition of a particle:

D1. An empirical individual *a* will be called a particle (represents a particle) in a temporal interval *t* if and only if there is a possibility of a spatial structure A with respect to methods of establishing spatial order α such that at the time *t* the individual *a* is inside A with respect to α, and *a* is in contact with the boundary points of A with respect to methods of establishing order that are included in α.

Let us also adopt the following definition of a wave:

D2. An empirical individual *a* will be called a wave (represents a wave) in a temporal interval *t* if and only if *a* is a process in *t* such that $x = f(y)$, where *a* is a variable for states of *a* in *t*, *y* is a variable for the time moments in *t*, and *f* is a certain periodic function.

The definition of particle does not contain any reference to a process and its periodicity. But this does not mean that a process is negated here; in other words, it does not follow from D1 that *a* cannot be a periodic process. The definition of the wave contains no reference whatsoever to spatial localization. But this does not mean that it is negated here, i.e., it does not follow from D2 that *a* cannot be spatially localized. Definitions D1 and D2 are constructed on different linguistic planes. They employ different sets of linguistic means that do not negate each other. Thus, according to D1 and D2 the assertion 'An individual represents a particle and a wave' is not a logical contradiction.

A different aspect of this is that it is not always practically possible and expedient to consider individuals as particles and waves at the same time. For those individuals the dimensions of which are close to the minimal, and in the intervals of time close to the minimal, it is practically possible and appropriate. And there is nothing strange in the fact that microscopic objects manifest themselves as particles and waves at the same time. Another circumstance looks more strange, namely, that macro-objects cannot be conveniently considered as such, that there are certain objects that are not particles and waves at the same time. But this strangeness is of a practical, not of a logical nature.

The assertion 'An individual is a particle and a wave' appears contradictory only inasmuch as that the meaning of the terms 'wave' and 'particle' implicitly includes assumption A. It follows from logical explication of these terms that assumption A is a foreign element in their definition. In reality, acceptance of A is more strange than its absence (not its negation, but precisely its absence).

The question whether all or not all empirical individuals are particles (waves) in the time interval of their existence is an extralogical one. From the point of view of logic, only the following is meaningful. The extralogical hypotheses 'Every empirical individual is a particle' and 'Every empirical individual is a wave' are logically consistent when considered both separately and together. Assertions concerning the existence of empirical individuals representing particles and of those representing waves are also extralogical. However, the latter can be verified in experiment while the former can be neither confirmed nor disproved experimentally. It is of interest that in experiment we can neither confirm nor disprove the logically consistent (but also extralogical) assertions 'Some empirical individuals in principle (under any circumstances) cannot be regarded as particles (as waves)'.

42. The Trajectory

A trajectory of the motion of an individual a in a temporal interval t is an ordered series of spatial locations of a such that $x = f(y)$, where x is a variable for spatial locations of a, and y is that for the time moments in t.

Let x^1 be a spatial location of the individual a at a time t^1, and x^2 a spatial location of a at a time t^2 after t^1. The interval between t^1 and t^2 is not smaller than the minimal. Assume that it is either minimal or close to it. The observer, in order to record the spatial location of an individual, must spend a time which is not less than the minimal, since observation is also an empirical event. Suppose that this time is either minimal or close to it. Recording the location of a at t^1 requires a time α. During this time a moves over from x^1 to x^2. Later (after t^1) the observer who starts to establish the location of a can no longer register it at x^2. Thus, when the motion of individuals is observed at distances and in time intervals that are close to the minimal, it is no longer possible to establish the dependence of their spatial location on time, i.e., their trajectory. The observer can determine just the final results of the individual motion during the time, and in the spatial regions, that are far from minimal. Therefore, the notion of trajectory loses practical significance.

It does not follow from the above that micro-objects cannot move along certain trajectories. The only result obtained is that it is equally meaningless to ascribe to micro-objects a capability to move along definite trajectories and to reject this capability. This kind of assertion can be neither proved nor disproved; they are unverifiable.

Further it does not follow from the above that we cannot define the term 'trajectory' for micro-objects in an appropriate way, so that assertions concerning micro-objects and containing this term would become verifiable. This can be done by describing the results of the micro-objects' action on macro-objects (devices).

43. The Part and the Whole

The terms 'part' and 'consisting of parts' ('whole') have multiple meanings. Consider two cases of their usage which are different in principle and which are frequently confused (with paradoxical consequences). The first case: and object a is regarded as a part of an object b (and b contains a

as a part) when (1) b represents a cluster of objects, a is included in this cluster, which in addition includes at least one other object; (2) a is a characteristic substructure of the structure b. In both instances actual or potential decomposability into objects containing a is implied (in the second case potential decomposability is assumed). At the same time an assumption A is quite naturally accepted: the dimensions (spatial or temporal) of b are greater than those of a ('a part is smaller than the whole'). The second case of usage of these terms corresponds to situations when b is decomposed into objects $a^1, \ldots, a^n$ or consists of their unification. An example of this case is a molecule of water which consists of hydrogen and oxygen atoms. We have in mind here, instead of the possibility in principle to discover the presence of $a^1, \ldots, a^n$ in b, the result of the unification of $a^1, \ldots, a^n$ or the result of decomposition of b. It is these cases that provide rich material for the criticism of 'common sense' and 'conventional' logic (e.g., the coalescence of two drops of mercury results in one drop of mercury, so that the rule $1 + 1 = 2$, as it were, has no universal meaning).

Consider now empirical individuals, the dimensions and the duration of existence of which are close to minimal. If b is of minimal dimensions (length or duration) and one accepts A, then a^i cannot be a part of b, since the dimensions of a^i cannot be smaller than the minimal. Thus, one has either to agree with the fact that the concepts of part and whole are not applicable to micro-objects, the dimensions of which are assumed to be close to the minimal, or to reject the assumption A as an illegitimate element of the definition of terms 'part' and 'whole'.

If the definition requires the presence of a particular assumption concerning the relation between part and whole, a weaker assumption B seems more appropriate: if a is a part of b, a's dimensions do not exceed those of b. The structure of individuals the dimensions of which are close to the minimal, however, cannot be visualized without destroying their wholeness. Here we observe only the product of their decomposition. And the individuals that have been formed as a result of such decomposition have dimensions, in their turn, that are not smaller than the minimal. Thus, if $a^1, \ldots, a^n$ represent the products of decomposition of an individual b which has the minimal dimensions, then the dimensions of all $a^1, \ldots, a^n$ are not smaller than those of b. If $a^1, \ldots, a^n$ are regarded as parts of b, under the conditions of the corresponding agreement (in particular, rejecting A) there is no logical inconsistency in it, although the fact of coinciding dimensions of parts and whole appears paradoxical. If this evokes a protest, however, we should simply stop applying the terms 'part' and 'whole' to the relation between $a^1, \ldots, a^n$ and b. After all, it is people who should rule words, and not vice versa.

The expressions 'divisible' and 'indivisible' are also ambiguous. First, divisibility and indivisibility are understood as, respectively, the possibility and the impossibility of discovering in the individual at least two discernible parts, the dimensions of which are smaller than its dimensions. Second, we have in mind, correspondingly, the possibility and the impossibility of dividing (decomposing, destroying) a given individual in such a way that at least two discernible individuals that are different from it are formed. It is only in the first case that the dimensions of a divisible individual are equal to the sum of dimensions of parts into which it is divided. In the second case it is possible that the dimensions of the products of decomposition are greater than those of the decomposed individual. This pertains to macro-objects as well as to micro-objects.

If the dimensions of an individual are minimal or close to minimal, it is indivisible only in the first sense. In the second case, it is an extralogical question as to what the individual is. At any rate, there is no contradiction between the assumption of its divisibility in the second sense and that in the first sense. If a body a, characterized by minimal dimensions, is decomposed to form two or more bodies, the volume of the latter is greater than that of a. As the products of an individual's decay can exceed in dimensions the decomposing individual, we cannot apply the term 'part' to any individuals representing the products of decay of a given individual a, under the condition that an assumption 'the dimensions of a part do not exceed those of the individual the part of which it constitutes' is made. For instance, one cannot call any products of a bomb's explosion its parts (and linguistic practice intuitively follows this rule).

44. Forecasts

Logically substantiated forecasts concerning empirical states are constructed according to the scheme:

$$x^* \wedge y \rightarrow z^*,$$

where x^* is a set of statements x pertaining to one or another region of the universe A at a time t^1 (i.e., x^* is x at t^1), z^* is a set of statements z pertaining to the state of A at t^2 after t^1 (i.e., z^* is z at t^2), and y is a set of statements by virtue of which one obtains z^* from x^*, and without which it would be impossible to obtain z^* from x^*. The statement z^* represents a forecast with respect to A if it is obtained at t^1. We shall call x the empirical condition of the forecast z^*, and y its logical condition.

With respect to the logical condition y of the forecast z^*, one can say that it either directly represents a statement of the type $w \rightarrow v$ or there are rules in logic such that

$$\vdash y \leftrightarrow (w \rightarrow v),$$
$$y \dashv\vdash (w \rightarrow v).$$

In these expressions $w \rightarrow v$ is such a statement that, according to the rules of logic (in particular, according to the rules for substitution of terms in place of variables or the species terms in place of generic ones), a statement

$$x^* \rightarrow z^*$$

can be deduced from it. Both the latter and the former are statements pertaining to connection since it contains a temporal relation of empirical states.

As was already noted, determinism is also understood as a heuristic principle by virtue of which one can predict the state of A at t^2 with any degree of completeness and accuracy (approximation) if only the state of A at t^1 is known with a sufficient degree of accuracy. In the limiting case, if one knows the state of A at t^1 with absolute completeness and accuracy, it is possible to predict with the same degree of completeness and accuracy the state of A at t^2.

Note that the structure of the principle under consideration itself is $C \rightarrow D$. Thus, if we establish that C is wrong, the falsity of $C \rightarrow D$ would not follow from it in any way. The latter can be true when C is not-true and can be not-true when C is true. The assertion to the effect that it is impossible to know at a given time t^1 everything about A with absolute accuracy, which is true from the logical point of view, does not by itself represent evidence against the principle of determinism. In particular, the assertion 'If we know everything about A at t^1, we can accurately predict the state of A at t^2' cannot be confirmed, neither can it be disproved since it is impossible to know everything about A at t^1. However, considered as an exterme formulation of a certain heuristic principle, which is efficient only if accompanied by a number of stipulations and when applied in a limited number of cases, it contains no danger at all. In general, the principle of determinism even in such an extreme formulation (just as many other principles that it has became fashionable to 'disprove' because of the progress of science) deserves a more condescending attitude than the prevailing one.

The second circumstance that must be taken into account here consists in the fact that usually formulations of the principle of determinism refer only to the empirical condition of the forecast and omit any logical condition. Does this principle include an assumption of the unlimited possibility of determining the logical conditions of forecasts or not? Depending on the concrete answer to this question, quite different principles result.

Even if we accept the unlimited possibility of constructing statements $w \to v$ which form the logical condition of forecasts for individual states, the existence of additional statements leads to the following conclusion:

If the antecedents of the statements $A^1 \to B$ and $A^2 \to C$, A^1 and A^2 represent additional statements, then $A^1 \wedge A^2$ cannot be true, and it is practically impossible to make use of the unification of these conditional statements into $A^1 \wedge A^2 \to B \wedge C$.

The statements that make the forecasts possible, and can be reduced finally to the form $w \to v$, are constructed and demonstrate validity only under certain conditions. If the statements A^1 and A^2 in $A^1 \to B$ and $A^2 \to C$ are complementary, they are complementary under the same conditions. But in this case only one of them can be used. They are both applicable only under different conditions α and β which are incompatible, in other words, for them it is true that $\sim(\alpha \wedge \beta)$. Thus, either the conditions should be the same but $\sim(A^1 \wedge A^2)$ or they are different but $\sim(\alpha \wedge \beta)$.

Complementary statements represent a special case of the relation between statements for which $A^1 \to \sim A^2$ and the conditions of truth α and β of statements A^1 and A^2, respectively, also depend on $\alpha \to \sim \beta$. Therefore, the unification of $A^1 \to B$ and $A^2 \to C$ into $A^1 \wedge A^2 \to B \wedge C$ turns out to be practically useless. And since we are dealing here with empirical connections such that by definition $B \wedge C$ does not follow from $A^1 \wedge A^2$, this kind of complex statement cannot be true.

Thus, the inevitable consequence of the existence of complementary statements is the impossibility in a number of cases of constructing such true statements that could become a logical condition of forecasts.

45. The Hypothetical Ontology

The term 'formal ontology' means here a set of statements about empirical objects deducible from definitions of the linguistic expressions present in these statements. By 'hypothetical ontology' we mean a set of statements about empirical objects that cannot be logically deduced from the definitions of linguistic expressions; they are accepted instead as extra-logical assumptions or are deduced from such assumptions. The assertions of hypothetical ontology, just as those of formal ontology, are formulated in terms of logic.

Among the assertions of hypothetical ontology we can find general statements such as 'Every quantitative variation is a consequence of qualitative variations', 'Nature is continuous', 'All processes in nature do not terminate instantaneously (they have certain inertia)' (extrapolation is based on these facts), 'Everything occurs in jumps', 'All processes in nature sooner or later decay (terminate)', 'Every progress sooner or later reaches its limit', 'All objects in nature are ordered', 'Chaos prevails in nature', etc.

Let us return to the statement

$$(\forall a)(\exists b)(b < \alpha a),$$

where a and b are the variables of the states of the Universe in time. We considered it above not from the point of view of its truth or falsity, but exclusively from the point of view of its logical structure. We can also concentrate our attention on an opposed statement. 'The Universe has a beginning in time', the logically explicit form of which is as follows:

$$(\exists a)(\forall b)(b \geqslant \alpha a).$$

But the question of which one of these assertions is to be accepted and which one rejected, is not a logical question. In this connection we shall make just the following remark.

Since in the first assertion it is impossible to consider individually all states of the Universe in the past (their number, according to this very assertion, is limited), and in the second one it is impossible to indicate which concrete state of the Universe is the initial one, empirical cognition does not discriminate between the above two assertions. The acceptance of each of them separately does not by itself lead to any logical contradiction.

The situation with other general hypotheses about the Universe is similar: analysis of their logical structure and their explication should not be confused with their acceptance. Consider the following hypothesis: 'Space is non-empty always and everywhere'. In logical terms it can be explicated in the following manner:

$$(\forall s)(\forall t)(\exists a)\mathrm{E}ts(a),$$

where s is a spatial variable, t is the variable of time, a is the variable of the body. However, it is impossible to check to what extent it is true. And there are many such hypotheses: 'There is no pure duration', i.e.,

$$(\forall t)(\exists a)\mathrm{E}t(a),$$

where a is the variable of variations; 'If something has emerged, it sooner or later will be destroyed'; 'If something has an end, it must have a beginning', etc. The set of such hypotheses includes those of the infinity of the Universe in space and in time in the future, of the inertia of nature, etc.

Nonetheless there are cases when logic is competent to accept or reject this kind of assertions. Their number includes situations for which certain statements or their negations are deduced according to the rules of logic from the definitions (explicit or implicit) of terms occurring in these statements. There are also cases when logic is competent to say that one or another assertion is uncertain. And such cases were frequently discussed here in previous sections. There is no general criterion for distinguishing between logical assertions (verifiable in logic) and extralogical assertions (indifferent with respect to logical constructions). In every individual case, the problem should be solved separately, with the possibility of surprising and apparently paradoxical phenomena. Thus, we have seen that assertions about minimal dimensions and durations and maximal velocities represent logical assertions, while those about the finiteness and the infinity of the Universe are extralogical ones, although it seems [intuitively] that the opposite should be true.

Consider hypotheses concerning the existence of the beginning of the Universe in time, and the absence of such a beginning, in greater detail. They will be used as an example for formulating some important general logical principles and ideas pertaining to the logical analysis of this kind of linguistic construction in general.

First of all, we shall write these hypotheses in a logically explicit form. Assume that x and y are the variables of terms of empirical events, and t is a method of establishing the temporal order of events. The hypothesis 'The Universe has no beginning in time' will be written as follows:

(1) $(\forall x)(\exists y)(x > ty)$,

(i.e., for any event there will be another one which occurred before it). The hypothesis 'The Universe has a beginning in time' will take on the following form:

(2) $(\exists x)(\forall y)(x \leqslant ty)$,

(i.e., there is an event such that all other events either have occurred either simultaneously with it or after it).

From (1) follows $\sim(2)$, and from (2) follows $\sim(1)$. But we should not consider one of (1) and (2) simply as negation of the other, since a third case is possible, namely,

(3) $\sim(1) \wedge \sim(2)$,

which corresponds to uncertainties (there may be several versions here). In order not to make discussion too complicated, we do not consider t as a variable nor introduce quantifiers for such a variable. If this is done, the number of versions will increase, since then we would have to take combination with $(\forall t)(1)$, $(\exists t)(1)$, $(\forall t)(2)$, $(\exists t)(2)$ and their negations into account. Thus, the above hypotheses have multiple meanings, with the reason for this being not just the ambiguity of the word 'beginning' (see above) but the variations of their explicit record as well. Our discussion will be limited only to versions (1), (2) and (3).

Note first of all that these hypotheses represent extralogical assertions. This can be demonstrated through conventional means of logic when one proves that each of $\vdash(1)$, $\vdash(2)$ and $\vdash(3)$ depend neither on axioms nor on the rules of inference of logic. Moreover, these hypotheses are logically realizable, in other words, the assertions $\vdash\sim(1)$, $\vdash\sim(2)$ and $\vdash\sim(3)$ are not provable in the framework of logic. The acceptance of any of $\vdash(1)$, $\vdash(2)$, $\vdash(3)$ separately does not by itself lead to contradiction. In particular, if $\vdash(1)$ is accepted, in the resulting system assertions $\vdash(2)$ and $\vdash(3)$ will not be provable; if $\vdash(2)$ is accepted, then $\vdash(1)$ and $\vdash(3)$ will not be provable; if one accepts (3), then $\vdash(1)$ and $\vdash(3)$ will not be provable. The rules of logic exclude the acceptance of any pair of them at the same time. If it is possible to derive from a certain set of extralogical assertions w the conjunction of two of them, then w is logically contradictory. And these hypotheses cannot be checked empirically.

It is possible to prove the consistency of a system obtained through addition to logic of $\vdash(1)$ if we determine a suitable semantic interpretation of formulas such as $\vdash\alpha(a > \beta b)$, where α is a certain sequence of quantifiers (it may be empty). This can be done, in particular, in the following way:

if a and b are different, then $a > \beta b$ can be ascribed a value v; if $a > \beta b$ can be ascribed the value v, then $\vdash (\exists b)(a > \beta b)$ has a value v. These semantic rules do not affect the assertions of logic. The consistency of a system resulting from adding $\vdash$(2) or $\vdash$(3) to logic can be proved in a similar manner. It follows from consistency of such systems that in the system with $\vdash$(1) assertions $\vdash$(2) and $\vdash$(3) are unprovable; in the system with $\vdash$(2) assertions $\vdash$(1) and $\vdash$(3) are unprovable; in the system with $\vdash$(3) one cannot prove $\vdash$(1) and $\vdash$(2). Thus, none of the hypotheses can be given logical preference. The well-known antinomies of Kant pertaining to space and time result simply from poor logical processing of terminology, confusion of different concepts, and implicit assumptions.

The spatial infinity of the Universe presents a quite similar situation. In this case the acceptance of one of (at least) three mutually excluding hypotheses also does not lead to the necessity of accepting the other, and none of them can be given logical preference.

A similar case is presented by the following hypotheses: (1) 'The number of individuals of any cluster of empirical individuals at any given time is finite'; (2) internal negation of (1); (3) conjunction of external negations of (1) and (2). The hypotheses 'The number of empirical individuals in the Universe at any given time is finite', 'The number of empirical individuals of any class at any given time is finite' represent a special case of hypotheses (1).

Hypotheses 'Nothing emerges from nothing' and 'Something does not turn into nothing' can be explicated in different ways, in particular, as follows:

$$(1) \qquad (\forall a)(\exists b)((\neg E(a) \Rightarrow E(a)) \rightarrow (b \Rightarrow a)),$$
$$(2) \qquad (\forall a)(\exists b)((E(a) \Rightarrow \neg E(a)) \rightarrow (a \Rightarrow b)).$$

From these hypotheses we can obtain logical consequences, such as

$$(\forall a)(\exists b)((\neg E(a) \Rightarrow E(a)) \rightarrow (E(b) \Rightarrow \neg E(b)),$$
$$(\forall a)(\exists b)((\neg Et^1(a) \Rightarrow Et^2(a)) \rightarrow Et^1(b)),$$
$$(\forall a)(\exists b)((E(a) \Rightarrow \neg E(a)) \rightarrow (\neg E(b) \Rightarrow E(b)),$$
$$(\forall a)(\exists b)((Et^1(a) \Rightarrow \neg Et^2(a)) \rightarrow Et^2(b)).$$

This example provides evidence of the fact that, when this kind of hypotheses are studied, it is possible to find a certain number of such hypotheses that allow us to derive, according to the rules of logic, other hypotheses (with a certain degree of completeness). This deducibility is possible inasmuch as these hypotheses are indeed general (do not imply terms of specialized sciences) and yield to explication within the framework of the language of logic.

It will be also interesting to consider the above hypotheses from the following point of view. They are meaningful not for arbitrary empirical objects. Consider, for instance, a relation '*a* is heavier than *b*' and eliminate *a*. The relation between *a* and *b* ceases to exist; it does not turn, however, into anything else. Or, we can create objects *a* and *b* and establish a certain spatial relation R between them; then we form a relation *a*R*b*. But it would be difficult to specify what exactly has turned into this relation, since the meaning of all the linguistic expressions appearing here prevents it. And the limitation of the meaningfulness of these hypotheses by certain definite logical types of objects constitutes a part of the implicit definition of the latter. In this case to say that the above hypotheses are meaningful only for empirical bodies is equivalent to tacit adoption, as part of the implicit definition of the empirical body, the above assertions (1) and (2) with the following limitation imposed on the variables; *a* and *b* are variables for terms of empirical bodies.

It is within the sphere of concrete analysis of individual hypotheses to establish to what extent the above hypotheses can turn into implicit definitions of certain logical types of objects.

The hypotheses under consideration are divided into general types that are valid for any empirical individual and special types that hold only for a certain type of empirical individuals. We have already given examples of general hypotheses. Let us cite a few more such examples which are also interesting in themselves.

H^1. If an empirical individual has emerged, it will disappear (becomes destroyed, will cease to exist).

H^2. If an empirical individual has disappeared, it had appeared.

H^3. For any empirical body (variation) and every way of establishing spatial (temporal) order there is an empirical body (variation) which is in contact with it with respect to this method.

H^4. There is never and nowhere in the Universe an absolute emptiness (i.e., at any time and in every spatial domain there exists a certain empirical individual).

It follows from H^4 that if the spatial domain has minimal dimensions, the individual existing in it also has minimal dimensions, so that the space is always completely filled with empirical individuals.

H^5. There is never and nowhere in the world an absolute rest (i.e., at any time in every spatial domain there is an empirical variation).

It follows from H^5, if the temporal interval is minimal, the variation occurs also during the minimal time, so that time is completely filled with empirical variations.

H^6. At any time and in every spatial domain there are such empirical individuals that one of them is transforming into the other.

H^5 follows from H^6 (but not vice versa). It also follows from H^6 that something does not turn into nothing and something cannot emerge from nothing.

H^7. If an empirical individual has a magnitude x with respect to a certain attribute, and after some time it has a magnitude y with respect to the same attribute the time interval being no less than the minimal), the remainder of x and y is not equal to zero.

These examples demonstrate that there is a logical connection between hypotheses of this kind. Thus, it would be of interest to find (as small as possible) a set of this kind of hypotheses from which the rest of them can be deduced or to establish the logical impossibility of such minimization of hypothetical ontology.

$$(H^8) \qquad E(a) \rightarrow (\exists s)(\exists t)\, Ets(a)$$

(i.e., if the individual exists, it exists at a certain time and in a certain spatial domain).

$$(H^9) \qquad E(a) \wedge \neg Es^1(a) \rightarrow (\exists s^2) Es^2(a),$$
$$(H^{10}) \qquad E(a) \wedge \neg Et^1(a) \rightarrow (\exists t^2) Et^2(a),$$

where y is 't^1 and t^2 are different time moments'.

$$(H^{11}) \qquad (x \rightarrow (Rx)y) \rightarrow (\exists \downarrow z)(\exists R^1)(\exists R^2)(z \rightarrow (R^1 z)x) \wedge (z \rightarrow (R^2 z)y) \wedge \wedge ((\downarrow z R^1 \downarrow x) \wedge (\downarrow z R^2 \downarrow y) \rightarrow (\downarrow x R \downarrow y)),$$

where R^1 and R^2 are variables for the relations of order.

H^{12}. The action of one individual on another occurs in quanta (in a discontinuous way).

H^{13}. The interacting objects act on each other alternately.

H^{14}. An object can act only on a finite number of objects, and this number does not exceed a certain maximum.

H^{15}. An object can be subject to the action of only a finite number of objects, this number being not greater than a certain maximum.

H^{16}. Interactions are subject to decay.

H^{17}. There are independent series of interactions.

An example of special hypotheses is provided by those pertaining to empirical points. Empirical points, lines, surfaces and bodies represent empirically existing individuals. They are characterized by certain properties that are fixed by extralogical assertions. Some of these assertions can be formulated completely in the language of logic (i.e., they are constructed of terms, definable in the frame of logic) and are of hypothetical nature. Consider a few cases of such assertions.

H_1. At any time and in every spatial domain there is EP (the non-emptiness hypothesis.

Since H_1 pertains also to the spatial domain of minimal dimensions, it follows from H_1 that space is everywhere and always filled with EP's.

H_2. The duration of existence of EP is close to the minimal.

H_3. For any two EP's there is a method of establishing the spatial order such that the distance between these EP's with respect to this method varies by a finite amount.

H_4. If an EP is eliminated, at least one other EP emerges as a result of it. And the emergence of EP is a consequence of the previous destruction of another EP. The time between the destruction of an EP and the resulting appearance of new EP's is close to the minimal.

Interesting conclusions can be drawn from the above hypotheses, e.g., the following: (1) The time of existence of every EL is either minimal or close to it; (2) The time of existence of every EF is either minimal or close to it; (3) No EP can return to the previous state. It follows from (3), that it is impossible for the Universe to return to the previous state. From (1) and (2) we can obtain the result that every absolute reference frame for the motion of bodies can exist only during minimal or close to the minimal time. Note incidentally that an opinion to the effect that such a system is logically impossible is wrong. There are an infinite number of absolute reference frames in the Universe.

For some cases, however, such a system is practically impossible because of the small value of the time of existence. But it is feasible in those situations when the quantities that are close to the minimal are sufficiently large from the practical point of view.

Finally, the above hypotheses are divided into those heuristically neutral (like, for instance, the hypothesis of the infinity of the Universe in space) and those heuristically effective (for example, the hypothesis according to which something cannot emerge from nothing and something does not turn into nothing; this hypothesis in some way generalizes the physical conservation principles).

Since the hypotheses under consideration can neither be confirmed nor disproved experimentally, while their acceptance is based on certain goals of cognition, they may be called heuristic hypotheses, and the system of such hypotheses is the heuristic ontology.

46. General Assertions Concerning the Universe and Physical Assumptions

There is a logical connection between the fundamental assumptions of physics and general assertions about the Universe; this connection manifests itself only when the corresponding linguistic expressions are logically explicated. Let us mention the most important features of this connection.

Consider, for instance, a physical assumption A: 'A body preserves its state of rectilinear uniform motion until external forces cause it to leave this state'. It can be partially explicated as follows: 'If a body is moving and no forces are acting on it (X), its motion is rectilinear and uniform (Y), with $(\forall a)(\exists b)(b > \alpha a)$', where a and b are variables for the spatial locations of the body, and α is a certain method of establishing the spatial order. From A it logically follows that, assertion B: 'If X, then $(\forall a)(\exists y)(y > \alpha a)$', where y is a variable for empirical individuals.

Further, let us make an assumption of the spatial finiteness of the Universe, explicated in such a way that from it one can obtain an assertion $(\exists x)(\forall y)(x \geqslant \alpha y)$, where x and y are variables for empirical individuals. From the latter we can obtain $\sim(\forall a)(\exists y)(y > \alpha a)$, where a is a variable for the spatial locations of the body (an assertion C). From B and C follows $\sim$X. Thus, a consequence of the assumption of the finiteness of the Universe in space is the non-truth of X, which excludes the possibility of using A for obtaining the true consequences.

It does not follow from the above that it would be in the interest of physics now to make an assumption of the infinity of the Universe in space. From what has been said, follows only that if we accept the law of inertia for motion, we should not assume the finiteness of the Universe in space, since this assumption renders the law practically useless. However, there is no logical contradiction here as yet. But if, nevertheless, we have to choose one of the two above assertions concerning the Universe, we obviously should prefer that of the infinity of the Universe in space.

It is possible to show that the assumptions 'Nothing emerges from nothing' and 'Something does not turn into nothing' are logically imcompatible with the assumptions of the finiteness of the Universe in time in the past and, correspondingly, in the future. It does not follow from this that we should assume the infinity of the Universe in the past and in the future. But there is a more rigid logical connection here than in the above, namely, a logical contradiction between assumptions. And this, again, can serve as a basis for preferring some hypotheses to others.

Further, there are physical assumptions from the explication of which we can derive an assertion A to the effect that every empirical individual has continuous surroundings (is immersed in a continuous medium) consisting of empirical individuals. Let us explicate A in the following way: $(\forall\alpha)(\forall a)(\exists b)((b > \alpha a) \wedge (a \parallel \alpha b))$, where a and b are variables for empirical individuals, and α is a variable for the methods of establishing the spatial order. From A follow assertions B and C, respectively: $(\forall\alpha)(\forall a)(\exists b)(b > \alpha a)$ and $(\forall\alpha)(\forall a)(\exists b)(a \parallel \alpha b)$. The first of them represents the explication of the assertion of the infinity of the Universe in space, while the second, of the assertion of the absence of absolute emptiness. Thus, the assertion of the infinity of the Universe in space can be obtained as a consequence of certain physical hypotheses. But the reverse is not true. Thus, from B and C, A does not follow. For time, a similar argument is valid.

Finally, let us consider a fact that suggests a more radical position on the question of the relation between formal and hypothetical ontology. Every extralogical assumption concerning empirical individuals, when explicated logically, has the form $(\forall a)x$, where a is a variable for empirical individuals in general or of their particular types (processes, clusters, etc.). A question arises: is it not possible to include in the set of values of a, such expressions as a $\downarrow \sim x$? If this is possible, we obtain by the rules of logic the result that x will be true with respect to such individuals a for which $\sim x$ is true (for example, we obtain that the individuals having no causes have causes). Therefore, assuming $(\forall a)x$, in order to avoid the possibility of logical contradictions, we should forbid substitution of expressions of the form $a\downarrow \sim x$ in place of a; this is equivalent to adopting the assertions $\sim (a\downarrow \sim x \rightharpoondown a)$ and $\sim (a\downarrow \sim x \in \mathrm{K}a)$. For instance, if we make an assumption that every existing expirical individual has a cause, we should also make an assumption according to which an object denoted by the term 'the existing empirical individual that has no cause' is not included in the class of existing empirical individuals. And this is equivalent to the operation when $(\forall a)x$ is simply adopted as a part of the implicit definition of terms 'an empirical individual', 'a process', 'a cluster', etc. Thus, every hypothetical ontology can in principle be presented as a system of definitions for certain types of empirical individuals and their attributes.

47. Generalizations of the Results of Science

Extreme [or philosophical] generalizations of the results of science are realized through linguistic expressions which can be explicated by the means of logic. Considered from a logical point of view, they are but assumptions with respect to empirical objects, no matter how many examples we cite in order to confirm them and how great the social prestige of these examples. These generalizations become provable if the linguistic expressions are defined in a suitable way; the generalizations are then obtained from such definitions as logical consequences. The construction of definitions of linguistic expressions within the framework of logic is, however, completely independent of the scientific results. Thus, the greatest achievements of physics (and any other science) do not contribute anything to the generalized picture of the Universe obtained here that would not be observable in a conventional experiment. And this is due to the properties of the very logical means of constructing linguistic expressions. The development of science provides stimuli for work in this direction and draws the investigator's attention to the facts that lie on the surface but have remained out of the sphere of the scholar's attention. Therefore, the most extreme generalizations of scientific results (if their meaning is established in a sufficiently rigorous and clear manner) turn out to be but banal truths, examples of which can be found in everyday life without any scientific equipment. They are not banal only if considered from the point of view of their ambiguity and uncertainty, i.e., as ideological rather than scientific phenomena. And it is only the question of provability of such generalizations which is a non-trivial question of the logical processing of the language.

APPENDIX

I. Epistemic Expressions

Those who deal with propositions we shall call propositioners. Assume that *a* is the name of the propositioner (a term denoting the propositioner), and *x* is a proposition. Every proposition represents a special perceptible (visible, audible, etc.) object and possesses a meaning and a truth value. We classify as epistemic propositions only those propositions that pertain to the relation of the propositioner *a* to the proposition *x* in these three aspects. These are the following propositions.

We shall say that a propositioner *a* is familiar with a proposition *x* as a perceived object if she can either pronounce or write *x*, has heard or seen *x*, remembers it, etc. The propositions recording this knowledge will be schematically written as '*a* has *x*' (or '*a* possesses *x*') and called the propositions of possession. The possession of a proposition represents a certain capability or a certain state of the propositioner. A propositioner can possess a proposition without knowing its meaning and truth value.

We shall say that a propositioner *a* is familiar with the meaning of a proposition *x* (understands the meaning of *x*) if she knows the value of all terms occurring in *x*, and the properties of all logical operators occurring in *x*. Obviously, there may be various degrees and forms of understanding the meaning of a proposition depending on the level and the nature of education, the speech skills, and so on, which is of no importance here. For us it would be sufficient to assume that if a propositioner is familiar (in some form and to some degree acceptable among a given circle of people) with the value of terms and the properties of logical operators occurring in *x*, then she is familiar with the meaning of *x* (understands *x*) from the point of view of this circle of people. In order to understand the meaning of *x*, one has to possess this proposition in one or another way. The propositions recording the fact that *a* knows the meaning of the proposition *x* will be schematically written as '*a* understands *x*' and called the propositions of understanding.

In everyday speech the expression '*a* does not understand *x*' is ambiguous. In one sense it means that *a* possesses a proposition *x* but does not know the value of certain terms or properties of some logical operators occurring in *x*. In another sense it means that it would be wrong to say that *a* understands *x*. And such negation can result not only from what has been said above but also from the fact that *a* does not possess *x* at all. Thus, it is necessary to distinguish between two kinds of negation here.

If a propositioner *a* believes (thinks, states, etc.) that a proposition *x* is true (or if the situation is indeed such as stated in *x*), we shall say that *a* recognizes (or accepts) *x*. The propositions in which this fact is recorded will be called propositions of acceptance (recognition) and schematically written as '*a* accepts (or recognizes) *x*'.

With the propositions of acceptance, one does not imply that the accepted proposition as true. A propositioner can recognize a proposition as true if in reality it is not. The acceptance of a proposition *x* by a propositioner *a* represents a certain action of *a*, in particular, such as: *a* says to other persons or to herself that she recognizes *x* (considers it as true); in response to the question whether *a* accepts *x* or not she answers 'yes', nods her head, and so on. It is in the propositions of acceptance that this kind of the propositioner's actions are recorded. These actions are diverse. We shall agree that *a* accepts *x* if and only if she realizes or can realize under certain conditions (for instance, on request, if compelled, voluntarily) a certain action which can be classified as that of the acceptance of propositions.

The negation of the statement '*a* accepts *x*' is also ambiguous. It is used actively, i.e., as '*a* accepts not-*x*' and passively, i.e., as '*a* does not perform any action which represents the acceptance of *x*'. We shall use it in the second sense.

Actions of acceptance of propositions are diverse. Thus, situations are possible (and they do occur in reality) when the propositioner accepts *x* in some way (e.g., she says to herself or to close acquaintances that she regards *x* as true) and refrains from doing so in another, in particular, she says before a certain group of persons that she does not accept *x*. According to our convention with respect to the passive meaning of the negation of the acceptance of a proposition, one should not see any contradiction in this fact. If the propositioner states that she does not accept *x*, in terms of our agreement this means only that she does not accept *x* in one way (through an official declaration of acceptance), which does not make it impossible that she accepts *x* in a different way. A propositioner will be in the state of logical contradiction only if she accepts *x* and at the same time accepts not-*x*.

In order to accept x as a proposition, the propositioner should understand it. Thus, here again there may be two possibilities for the negation of the statement 'a accepts x', namely, the following ones: (1) a understands x but performs no actions to accept it; (2) one cannot say that a accepts x for the reason which not only coincides with that given in (1), but also of not understanding x.

The propositions of acceptance are related to those of rejection and indifference. The former will be written as 'a rejects x'. They represent abbreviations or replacements of propositions of the form 'a does not accept 'not-x'. The latter we shall write as 'a is indifferent with respect to x'. They are abbreviations for such propositions as 'a does not accept x and does not accept not-x' or 'a does not accept and does not reject x'.

Finally, epistemic propositions include propositions to the effect that a propositioner a knows that the proposition x is true (a is informed of x; a knows that x). Such propositions will be called the informative propositions or the propositions of knowledge; they will be written schematically as 'a knows that x'.

In the case of informative propositions it is assumed that x is true, and a accepts it in some way (and, therefore, possesses the proposition x and understands it). A special case of such statements are propositions of the form 'a guesses that x'. 'a has heard that x', etc. In general, there are certain states of propositioners which are recorded in the informative propositions.

For the informative propositions it is also essential to distinguish between two kinds of negation. Without this, paradoxical consequences result. Thus, it is intuitively obvious that when they say 'a knows that x' the truth of x is implied. Quite similarly, one assumes that x is true when one says 'a does not know that x'. This logical association between statements can be expressed through the conditionality operator as follows: (1) if a knows that x, then x; (2) if a does not know that x, then x. By applying the rule of contra-position, we obtain from the first statement: (3) if not-x, then a does not know that x. And from assertions (2) and (3), according to the transitivity rule, follows that: (4) if not-x, then x. An assertion (5) if x, then x holds. From (4) and (5) we obtain: (6) if x or not-x, then x. And from (6), by virtue of the fact 'x or not-x' is logically true, it results that any proposition x is true.

But regardless of such paradoxes the following is obvious. The negation of the statement 'a knows that x' may mean that x is true but a does not accept it, and that the statement 'a knows that x' is wrong. In the second case the reason may be not just non-acceptance of x, but also the non-truth of x.

All of the above statements contain two subject-terms each (the term a and the term 'a proposition x' or 'the fact that x') and two-place predicates, respectively, of possession, understanding, acceptance, rejection indifference, and informative predicates. For the sake of brevity, we shall denote these predicates by letters A, B, C, D, E, F, respectively, and propositions on the whole by symbols A*amx*, B*amx*, C*amx*, D*amx*, F*amx*.

We would like to draw the reader's attention to the fact that the epistemic statements P*amx* contain not just the statements x, but the terms mx which are characterized by a certain peculiar feature. Consider a statement: 'a has said that x' representing the special case of the statement 'a has x'. Suppose that x is equivalent to y. If we adopt the rule of the substitution of equivalence, we obtain a statement 'a has said that y' which can be erroneous when the proposition 'a said that x' is true. It is impossible to avoid such paradoxes without recording the above peculiarity of terms. And this peculiarity consists in the fact that such terms contain x not as a proposition but as a physical (audible, visible) object. Hence an important conclusion of a general logical significance. It is not always correct to believe that a linguistic expression is a part of another one if and only if it represents a part of the latter. In each case it is necessary to determine accurately when a linguistic expression is an occurrence in another one as a linguistic expression of a given type.

The above pertains not only to epistemic expressions. Thus, from the fact that a certain given statement x is graphically longer than a statement y, it does not follow that x is longer than z even if y and z are equivalent, because the rule of equivalent substitution should be applied to the replacement of the occurrences of propositions as such, and not as physical objects.

The task of epistemic logic is to establish the rules for manipulating propositions with the above predicates, in other words, to define these predicates by its own methods. This should be done in correspondence with those intuitive considerations that have been partially presented above and partially will become clear from the very form of constructed calculi.

The definition of the properties of epistemic predicates that do not depend on the relation of the propositioners to the rules of logic and are valid for each propositioner separately is obtained from the system SE^1. This system is formed through the following additions to the logical systems discussed above. The symbols $\wedge$, $\vee$, $\urcorner$, $\sim$, $\vdash$ are, respectively, the operators of conjunction, disjunction, internal negation, external negation and the predicate of logical entailment. The expression $x \dashv\vdash y$ will be used as an abbreviation for two expressions $x \vdash y$ and $y \vdash x$.

Addition to the alphabet (or the proper alphabet of SE[1]);

(1) a list of variables for terms, denoting propositioners:

(2) A, B, C, D, E, F, – constant two-place epistemic predicates of possession, understanding, acceptance, rejection, indifference, information, respectively.

Addition to the definition of a predicate formula: if a is a variable for names of the propositioners, and x is a propositional formula, then Aax, Bax, Cax, Dax, Eax, Fax are predicate formulas.

The definition of the occurrence of a propositional formula in a propositional formula should take on the following form. A propositional formula x occurs in a propositional formula in the following and only in the following cases: (1) y occurs in $\sim y, y \wedge z, z \wedge y, y \vee z, z \vee y$; (2) $x^i i = 1, 2, \ldots, n)$ occurs in $x^1 \wedge x^2 \wedge \ldots \wedge x^n$ and in $x^1 \vee x^2 \vee \ldots \vee x^n$; (3) if y occurs in z, and z in v, then y occurs in v.

Additional axiomatic schemes:

(1) $\sim \mathrm{A}ax \vdash \urcorner \mathrm{A}ax$.
(2) $\mathrm{A}ax \vdash \mathrm{A}ay$, where y occurs in x.
(3) $\mathrm{B}ax \vdash \mathrm{B}ay$, where y occurs in x.
(4) $\mathrm{B}ax \vdash \mathrm{A}ax; \urcorner \mathrm{B}ax \vdash \mathrm{A}ax$.
(5) $\sim \mathrm{B}ax \vdash \sim \mathrm{A}ax \vee \urcorner \mathrm{B}ax$.
(6) $\mathrm{C}ax \vdash \mathrm{B}ax; \urcorner \mathrm{C}ax \vdash \mathrm{B}ax$.
(7) $\sim \mathrm{C}ax \vdash \sim \mathrm{B}ax \vee \urcorner \mathrm{C}ax$.
(8) $\mathrm{D}ax \dashv\vdash \mathrm{C}a \sim x$.
(9) $\urcorner \mathrm{D}ax \dashv\vdash \urcorner \mathrm{C}a \sim x$.
(10) $\mathrm{E}ax \dashv\vdash \urcorner \mathrm{C}ax \wedge \urcorner \mathrm{D}ax$.
(11) $\urcorner \mathrm{E}ax \dashv\vdash \mathrm{C}ax \vee \mathrm{D}ax$.
(12) $\mathrm{F}ax \dashv\vdash \mathrm{C}ax \wedge x$.
(13) $\urcorner \mathrm{F}ax \dashv\vdash \urcorner \mathrm{C}ax \wedge x$.

Axioms (1) mean that for predicate A, negations are not distinguished, since in the theory of predication the formula $\urcorner \mathrm{A}ax \vdash \sim \mathrm{A}ax$ is provable. The meaning of Axioms (8–13) is that predicates D, E, F are derived and can in general be eliminated from the language. The paradoxes discussed above do not result here, since the formulas $\sim \mathrm{F}ax \vdash \urcorner \mathrm{F}ax$ and $\sim \mathrm{F}ax \vdash x$ are unprovable. If $\urcorner \mathrm{C}ax$ is interpreted in the active sense, i.e., if the axioms

$\neg Cax \dashv\vdash Ca \sim x$ are adopted, the formula $\vdash \sim Eax$ will be provable, in other words, we obtain an assertion to the effect that the propositioner cannot be indifferent to any proposition. And this does not agree with intuition.

The relations of propositioners to the rules of logic are so diverse and individual that it is impossible to indicate here an element that would be common for all of them. In this case one can just give a definition of various types of propositioners.

If we assume that the propositioner knows and accepts certain rules of logic, the system of epistemic logic would not become richer than SE^1. The above assumption should be realized in the adoption of additional axioms that are valid for one or another type of the propositioners.

But an assumption that the propositioner knows and accepts the rules of logic implies that these rules are established and known by somebody. And if they are still not established at all, not accepted, etc.? The situation is resolved when a quite definite class of the rules of logic is implied. Below we shall give examples of the definition of the types of propositioners, in which the class of rules of logic is not limited by concrete logical calculi. The system SE^2 is formed by adding the following definitions to SE^1.

D1. A propositioner a will be called primitively logical if and only if the axiom schemes

(14) $Cax \vdash \sim Ca \sim x$,
(15) $Cax \wedge Cay \vdash Ca(x \wedge y)$,
(16) $Cax \dashv\vdash CaCax$,

are valid for her.

It can be derived from D^1 that if a is primitively logical, the following theorem schemes: $Fax \vdash \sim Fa \sim x$, $Fax \vdash FaCax$, $Fax \vdash CaFax$, $Fax \vdash FaFax$, etc. hold for it.

D2. A propositioner a will be called practically logical if and only if the axiom schemes (14–16) are valid for her, in addition to the following ones:

(17) $Ca(\vdash x) \vdash Cax$,
(18) $Ca(x \vdash y) \vdash (Cax \vdash Cay)$,

(with the expressions that contain the predicate of entailment being included in the number of the predicate formulas of the calculus SE^2).

As follows from D^2, if *a* is practically logical and accepts the rule $x \wedge y \vdash y$, the theorem schemes $\mathrm{F}ai^1\mathrm{F}b^1,\ldots,i^n\mathrm{F}b^n j\mathrm{F}cx\mathrm{F}ai^1\mathrm{F}b^1,\ldots,i^n\mathrm{F}b^n x$ will be valid for her; in the above schemes $n \geqslant 1$ and $i^1,\ldots,i^n,j$ stands for the presence or the absence of the internal negation (in arbitrary combinations). A special case is provided by the theorems $\mathrm{F}a\mathrm{F}bx \vdash \mathrm{F}ax$ and $\mathrm{F}a\urcorner\mathrm{F}bx \vdash \mathrm{F}ax$. The formulas $\vdash \sim\mathrm{F}a\urcorner\mathrm{F}ax$ and $\mathrm{F}a(x \vdash y) \vdash$ $\vdash (\mathrm{F}ax \vdash \mathrm{F}ay)$ will be also provable. If a rule $(\vdash x) \vdash x$ is adopted, the formulas $\mathrm{F}a(\vdash x) \vdash \mathrm{F}ax$ will be theorems.

D3. A propositioner *a* will be called logical in the abstract sense if and only if, besides the axiom schemes (14–16), the following axiom schemes:

(19) $(\vdash x) \vdash \mathrm{C}ax$,

(20) $(x \vdash y) \vdash (\mathrm{C}ax \vdash \mathrm{C}ay)$,

hold for her.

Obviously, an answer to the question as to when x and $x \vdash y$ are valid is up to the authors of logical systems.

Consider now the interrelations between two propositioners *a* and *b* such as follows. Assume that F*ax*. From it neither F*b*F*ax* nor F*b*F*ax* follow, which is obvious. Assume that F*b*F*ax*. It is also obvious that both F*a*F*b*F*ax* and ~F*a*F*b*F*ax* do not follow from it (i.e., *a* may and may not know that F*b*F*ax*). Suppose that F*a*F*b*F*ax*. Again, it is possible then that *b* knows about this, and it is possible that he (she) does not know it. Assume that F*b*F*a*F*b*F*ax*. But at further stages the intuitive clarity disappears. And a question arises: can one construct a statement F*awx* or F*bwx*, where *w* is a certain sequence of letters F, *a*, and *b*, such that the assertions F*awx* F*b*F*awx* or F*bwx* F*a*F*bwx* hold? Our argument is based on the fact that in such situations the propositioners sooner or later guess that one of them knows that which is known to the other in connection with her and the proposition. Thus, we assume that in communication between *a* and *b* with respect to *x* beginning from a certain moment, the ascribing of F*a* to F*bwx* and F*b* to F*awx* becomes practically meaningless.

Suppose that *nab* is an expression in which an expression F*a*F*b* is written *n* times in a row ($n \geqslant 1$), and [*nba*] is an expression in which *n* times in a row F*b*F*a* is written. For instance, F*a*[2*ba*]*x* is an expression F*a*F*b*F*a*F*b*F*ax*. Let us adopt the following definition realizing our assumption.

D4. A pair of propositioners a and b will be called a mutually quick-witted pair of nth degree if and only if the axioms:

(21) $$[(n+1)ab]x \wedge [(n+1)ba]x \vdash [(n+2)ab]x$$

hold for them.

If $n=1$, the Axioms (21) have the following form:

$$FaFbFaFbx \wedge FbFaFbFax \vdash FaFbFaFbFaFbx.$$

As follows from 21,

$$[(n+1)ab]x \wedge [(n+1)ba]x \vdash [(n+2)ba]x.$$

If $n=1$, this corollary has the form:

$$FaFbFaFbx \wedge FbFaFbFax \vdash FbFaFbFaFbFax.$$

For practically logical propositioners one can derive from (21) that

(22) $$Fa[(n+1)ba]x \vdash [(n+2)ba]x.$$

For $n=1$ this corollary has the form:

$$FaFbFaFbFax \vdash FbFaFbFaFbFax.$$

For practically logical propositioners it also follows from (21) that

$$CaCbCaCbx \wedge CbCaCbCax \wedge CaCbCax \wedge CbCaCbx \wedge \wedge CaCbx \wedge CbCax \wedge Cax \wedge Cbx \vdash CaCbCaCbCaCbx \wedge \wedge CbCaCbCaCbCax.$$

And vice versa; for the propositioners of the above type one can derive from these formulas expressions (21).

Definitions, similar to D^4, can be adopted for three and more (in general, for any n greater than two) propositioners. In particular, it can be done as follows.

D5. An n-tuple ($n \geqslant 2$) of propositioners $a^1, a^2, \ldots, a^n$ will be called a mutually quick-witted n-tuple of mth order ($m \geqslant 1$) if and only if the following axioms hold for them:

$$[(m+n-1)y^1]x \wedge \ldots \wedge [(m+n-1)y^k]x \vdash [(m+n)y^1]x,$$

where $y^1, \ldots, y^k$ represent all possible permutations of $a^1, a^2, \ldots, a^n$.

We have cited D^5 only to illustrate the general idea. It is not our concern to which extent the antecedent of the above formula can be minimized.

If one ignores the antecedents themselves of formulas (21) and (22) on the whole, the sets of corollaries deducible from them which are formed of the letters F, a, b and x only coincide; these are the expressions Fax, Fbx, $FaFbx$, $FbFax$, $FbFaFbx$, $FaFbFax$, $FaFbFaFbx$, $FbFaFbFax$. And in this sense those versions are equivalent.

II. On the Concept of Belief

The statements of belief 'N believes that X' represent a special case of epistemic statements. Their symbolic form will be

$$W(N, mX)$$

where W is the predicate of belief ('believes in').

What is the intuitive aspect of such statements (and predicates)? Assume that A is a statement 'N has neither proof nor experimental verification of X or in general knows or believes that they do not exist'; B is a statement 'N accepts X'; C is a statement 'N is free to accept or reject X'. The proposition $W(N, mX)$ is just an abbreviation for $A \wedge B \wedge C$, in other words,

$$W(N, mX) \dashv\vdash A \wedge B \wedge C.$$

Those are the only general properties of the predicate W. Other statements, pertaining to W, characterize just the believers themselves. For instance, N will be called a consistent believer if and only if $\vdash \sim (W(N, mX) \wedge W(N, m \sim X))$; N will be called an abstract logical believer if and only if the following is valid: if $x \vdash y$, then $W(N, mX) \vdash W(N, mY)$ and $W(N, m \sim Y) \vdash W(N, m \sim X)$; N will be called a practical logical believer if and only if for N knowing that $x \vdash y$ we have $W(N, mX) \vdash W(N, mY)$ and $W(N, m \sim Y) \vdash W(N, m \sim X)$.

III. On the Notion of Preference

The statements of preference in a logically complete and explicit form are such statements as 'N prefers and action X to other actions from a set of actions Y'. For the sake of brevity we shall use a symbolic form

$$\Pi(N, X, Y).$$

Their truncated and transformed versions are also used: 'N prefers X' (and in this case a certain Y is implied), 'X is preferable to other Y' (here a certain N is implied), 'X is preferable' (a certain Y and a certain N are implied). An expression of the form 'N prefers a statement X to others from Y' represents only a transformation of the expression 'N prefers the adoption (acceptance) of X to acceptance of other Y's'.

What is the intuitive meaning of the statements (and the predicate) of preference? They pertain first of all to actions. Other cases can be interpreted as derived, i.e., simply as abbreviations for the fundamental cases. For instance, a statement 'N prefers classical music' is sometimes just an abbreviation for 'N prefers to listen to classical music'. Further, these expressions imply the following conditions: A certain set of actions X is given. This set is decomposed into two non-empty and non-intersecting subsets Y and Z. In a special case Y or Z or each of them includes one action. Thus, the set X contains at least two actions. Note that since non-realization of an action can also be an action, the set X can consist of two such actions each of which being an opposite of the other ('to smoke' and 'not to smoke', 'go to a movie' and 'not to go to a movie', etc.).

The symbol O(N, X) will denote the expression 'N realizes X'. The meaning of the predicate O for individual actions is regarded as intuitively clear. If X is a positive action, $\tilde{X}$ is an action, negative with respect to it. It is defined as follows:

$$O(N, \tilde{X}) \dashv\vdash \sim O(N, X).$$

Hence

$$O(N, \tilde{\tilde{X}}) \dashv\vdash O(N, X)$$

In the case of sets of two or more actions the predicate O is defined through

$$O(N, X) \dashv\vdash (\exists a)(O(N, a) \wedge (a \in X)).$$

In this expression O(N, X) stands for 'At least one of the actions of the set X is realized'.

Further, actions included in X possess the following properties: if a certain action from a set Y is realized, then there is no realization for any of actions from Z, in other words,

$$\vdash \sim (O(N, Y) \wedge O(N, Z)).$$

Therefore,

$$\vdash O(N, Y) \wedge \sim O(N, Z) : O(N, Z) \wedge \sim O(N, Y) : \sim$$
$$\sim O(N, Y) \wedge \sim O(N, Z).$$

The case when

$$\sim O(N, Y) \wedge \sim O(N, Z)$$

is not excluded.

Assume that a certain being N is facing a situation when there is a possibility of choosing a certain action (actions) from X and realizing this action. She may choose nothing. If an action from Y (or from Z) is chosen, no action can be chosen from Z (respectively, from Y). For such singular case the predicate Π is defined in the following manner:

$$\Pi(N, Y, X) \dashv\vdash O(N, Y) \wedge (\exists Z)[(Y \subset X) \wedge (Z \subset X) \wedge \wedge (X \subset \supset Y \cup Z) \wedge (\exists a)(a \in Y) \wedge (\exists a)(a \in Z) \wedge \wedge (\forall a) \sim ((a \in Y) \wedge (a \in Z)) \wedge ((a \in Y) \wedge (b \in Z) \rightarrow \sim (O(N, a) \wedge \wedge O(N, b))),$$

where a and b are action variables, and Z is a variable for sets of actions.

If the number of cases when the being N has to choose and realize an action from X is greater than two, the predicate Π for such cases individually. In particular, if there is regularity in N's choosing actions from X, and if the number of cases of choice is sufficiently great, Π can be defined either through the number of cases of choice or through the ratio of the number of choices of actions from a given subset Y to a certain assumed number of cases of choice of actions X in general (in particular, as the probability of choice). For instance, if the number of cases of choice and realization of a certain action from Y equals α per each number β of the cases of choice and realization of actions from X, the following agreement can be made:

$$\Pi(N, Y, X) \dashv\vdash \left(\frac{\alpha}{\beta} > \frac{1}{2}\right).$$

One can define through Π other predicates of preference, such as

(1) $\square(N, (Y, Z), X) \dashv\vdash \sim \Pi(N, Y, X) \wedge \sim \Pi(N, Z, X).$

(2) $\square(N, Y, X) \dashv\vdash \sqcup(N, (Y, Y), X).$

In (1) we have given the definition of equal preference, and in (2) of indifference.

A special case is provided by the proposition of preference of propositions. Assume that iX is 'the acceptance of X'. Then there is a possibility of defining the types of preferring subjects. For instance, N will be called a consistently preferring subject if and only if

$$\Pi(\mathrm{N}, i\mathrm{X}, \mathrm{U}) \vdash \sim \Pi(\mathrm{N}, i \sim \mathrm{X}, \mathrm{U}).$$

And in general, the 'logic of preference' as well as 'logic of belief', is just a system of definitions of the preference predicates and the types of preferring persons.

IV. On the Logic of Norms and Questions

Each norm corresponds to a normative statement, i.e., a statement containing the name of action and the predicates 'forbidden', 'allowed' or a predicate derived with respect to them. For example, the norm 'No Smoking' corresponds to a statement 'It is forbidden to smoke here'. Thus, one may adopt the following rule: a norm B follows from a norm A if and only if from the normative statement corresponding to A follows the normative statement corresponding to B.

A question represents a linguistic expression consisting of two parts: (1) a part of assumed or possible statement; (2) a linguistic expression denoting a suggestion (request, order) to complement the first part so that a certain true statement results. The latter represents the correct answer to the question. Thus, the following rule can be adopted: from a question A follows a question B if and only if from the correct answer to A follows the correct answer to B.

Thus, the idea of generalizing the notion of logical entailment in such a way that it would be applicable to linguistic expressions which are not statements is by itself absurd. The only meaningful concept here is that of reduction of relations of such expressions to those of statements.

V. On the Logic of Estimates

Estimate statements have the form

$$\mathrm{PR}(a),$$

where a is the term of the estimated object (in particular, this may be a pair, a triplet and, in general, an n-tuple of two or more objects), P is the name of the attribute with respect to which the estimate a is constructed, R is an estimate expression such as 'good', 'bad', 'five points', etc. The logic of estimates establishes the specific logical properties of estimate expressions and nothing else. For instance, a statement 'better' possesses properties which are established in the logic of comparison. The logic of estimates should be concerned with that of its properties which is shared by it and other estimate expressions and is absent in non-estimate expressions (e.g., in 'higher').

The relation between P and PR is defined by a rule:

(1) $\vdash (\mathrm{PR} \rightharpoonup \mathrm{P})$.

It follows from (1) that

$$\vdash ((a \downarrow \mathrm{PR}) \rightharpoonup (a \downarrow \mathrm{P})),$$
$$\vdash \mathrm{PR}(a) \rightarrow \mathrm{P}(a),$$
$$\vdash \sim \mathrm{P}(a) \rightarrow \sim \mathrm{PR}(a).$$

For instance, a bad athlete is an athlete; if a person is not an athlete, one cannot say about her that she is a bad (good, etc.) athlete.

Estimates can be classified into subjective and objective. In the first case there are no common (for a given circle of persons) criteria of estimate. Thus, if a person N^1 expressed $\mathrm{PR}(a)$, another person has no criteria whatsoever to establish whether this is true or not. In the second case there are such criteria (with a certain degree of completeness, stability, clarity, etc.). Further, estimates can be divided into absolute and relative. The former are applied when the term a in $\mathrm{PR}(a)$ is not an n-tuple of two or more terms, while the second, when a is $(a^1, \ldots, a^n)$, where $n \geqslant 2$. Finally, we can distinguish between basic and derived estimates. The former cannot be deduced from other estimate statements, whereas the latter, on the contrary, are derived from other statements of the same kind. The derived estimates, in their own turn, can be classified into logically derived and actually derived. Let us give an example of logically derived estimates. From the statements 'a is better than b as an athlete' and 'b is better than c as an athlete' one can deduce a statement 'a is a better athlete than c'. If $\mathrm{PQ}(a)$ is regarded as an abbreviation for $\mathrm{PR}^1(a) \vee \mathrm{PR}^2(a)$, then the assertions $\mathrm{PR}^1(a) \rightarrow \mathrm{PQ}(a)$ and $\mathrm{PR}^2(a) \rightarrow \mathrm{PQ}(a)$ (where PQ is an estimate derived with respect to PR^1 and PR^2) will hold. Such cases are of no interest since they do not at all take into account the specificity of the estimate expressions. The following is an example of the actually derived estimate. It is known what marks the students of a given class have received in mathematics. Computing the arithmetic mean mark, we obtain the estimate of the class as a whole. It is exactly those cases that are interesting from the point of view of logic. In general, the task of the logic of estimates appears to us as that of establishing certain rules pertaining specifically to estimate expressions.

Let us consider the criteria of estimates by analogy with statement coordinates. We shall introduce a peculiar sort of coordinates that are characteristic of estimate statements only. If x is an estimate expression, and v is a term of the criteria of estimates, then $[xv]$ is also an estimate statement. For the estimate criteria the following rule holds:

$$(2) \qquad [xv^1] \wedge (v^1 \rightarrow v^2) \vdash [xv^2].$$

When the symbols of the estimate criteria are omitted, it is implied that they are identical for all estimate statements in given expression.

Estimate statements with different estimate criteria are logically independent, i.e.,

$$(3) \qquad \vdash \sim(v^1 \rightarrow v^2) \wedge \sim(v^2 \rightarrow v^1) \rightarrow \sim([xv^1] \rightarrow [yv^2]) \wedge \wedge \sim([yv^2] \rightarrow [xv^1]).$$

Different basic estimates are independent, i.e.,

$$(4) \qquad \vdash \sim(\mathrm{R}^1 \rightarrow \mathrm{R}^2) \wedge \sim(\mathrm{R}^2 \rightarrow \mathrm{R}^1) \rightarrow \sim(\mathrm{PR}^1(a) \rightarrow \mathrm{PR}^2(a)).$$

For instance, from the fact that a student has got an A does not follow that she has got a B.

Suppose that B represents an n-tuple of two or more individuals. Let $a \in \mathrm{B}$ denote that a is one of the individuals occurring in B (a belongs to B). If all individuals of B are estimated with respect to an attribute P, there will be (is possible) a method of estimate of B with respect to the same attribute P, in other words,

$$(5) \qquad \vdash ((\exists v)((a \in \mathrm{B}) \rightarrow (\exists \mathrm{R})\mathrm{PR}(a)) \rightarrow (\exists v)(\exists \mathrm{R})\mathrm{PR}(\mathrm{B}),$$

where R is a variable for the estimate expressions, and v is a variable for the estimate criteria.

For example, if all students of a class have got marks for a given subject, the class on the whole can be ascribed a mark (e.g., the arithmetic mean of the individual marks).

The estimate of an n-tuple of individuals depends solely on those of individuals occurring in it, i.e., if the estimate criteria for B are given, and if one knows the estimates of all individuals occurring in B, this is sufficient for obtaining the estimate of B (rule 6).

Various truth estimates of individuals are equivalent, i.e.,

$$(7) \qquad \vdash [\mathrm{PR}^1(a)v^1] \leftrightarrow [\mathrm{PR}^2(a)v^2].$$

We have outlined only a possible orientation of the logic of estimates which, as it seems to us, is more interesting and less trival than the tradition now being formed in this field of logic.

VI. On the Theory of Proof

By the words 'the theory of proof' we mean a section of logic which specifies the properties of the predicate 'proved' (or 'provable'). Such expression as 'A statement x is provable' will be written symbolically as

$$\vdash x,$$

where $\vdash$ is the predicate of provability. We do not introduce here a special symbol different from the predicate of logical truth because of the following considerations. Assertions 'A statement x is provable (proved)' are accepted when and only when (1) x is a theorem of logic; (2) x is a definition, a part of a definition or a consequence from a definition, and in this case x represents a theorem of logic; (3) x is an assumption, and in this case an assumption 'Let x be provable' does not differ in any respect from the assumption 'Let x be logically true'. To put it briefly, we do not permit any source of provability other than the definitions of terms and logical operators and consequences of those. Below we shall present some additions to our logical theory which form a fragment of proof theories. We believe that an absolutely complete theory of proof is impossible in principle, just as an absolutely complete logic in general. In what follows we shall also demonstrate (with an example of the fragment discussed and later with that of the theory of complete induction) that the theory of proof contributes something new to the theory of inference in general.

An addition to the definition of the propositional formula:

(1) if x is a propositional formula, then $(\vdash x)$ is a propositional formula;

(2) if x and y are propositional formulas, then $(x \vdash y)$ is a propositional formula.

Additional axioms (or axiomatic schemes):

(1) $(\vdash x) \vdash x$.

(2) $(\vdash x) \vdash (\vdash (\vdash x))$.

(3) $\sim (\vdash x) \vdash (\vdash \sim (\vdash x))$.

(4) $(\vdash x) \wedge (\vdash y) \vdash (\vdash (x \wedge y))$.

(5) $(x \vdash y) \vdash (\vdash (x \vdash y))$.

(6) $\sim (x \vdash y) \vdash (\vdash \sim (x \vdash y))$.

(7) $(x \vdash y) \vdash ((\vdash x) \vdash (\vdash y))$.

(8) $(x \vdash y) \vdash (\vdash (\vdash (x \rightarrow y))$.

(9) $(x \wedge z \vdash y) \wedge \sim (z \vdash y) \wedge (\vdash z) \vdash (\vdash (x \rightarrow y))$.

(10) $(\vdash \sim (x \rightarrow y)) \vdash \sim (x \vdash y)$.

(11) $\sim (\vdash (x \rightarrow y)) \vdash (\vdash \sim (x \rightarrow y))$.

(12) $(\vdash \sim (x \rightarrow y)) \vdash \sim (\vdash (x \rightarrow y))$.

(13) $(\exists \mathrm{Q})(((\vdash x) \rightarrow \mathrm{Q}(mx)) \wedge (\exists a)(\mathrm{Q}(m\mathrm{A}) \wedge \mathrm{Q}(m \sim \mathrm{B})) \vdash$
$\vdash (\vdash \sim (\mathrm{A} \rightarrow \mathrm{B}))$.

(14) $(\exists \mathrm{Q})(((\vdash x) \rightarrow \mathrm{Q}(mx)) \wedge (\exists a)(\mathrm{Q}(m\mathrm{A}) \wedge \sim \mathrm{Q}(m\mathrm{B})) \vdash$
$\vdash (\vdash \sim (\mathrm{A} \rightarrow \mathrm{B}))$.

(15) $\vdash (\exists a) \sim x \rightarrow \sim (\vdash x)$, where a freely occurs in x.

(16) $\sim (x + y)$, if at least one variable occurs in y and does not occur in x.

(17) $\vdash \sim (\vdash x) \rightarrow \sim (\vdash (x \wedge y))$.

(18) $\vdash \sim (x + y) \rightarrow \sim (x \vdash y \wedge z)$.

(19) $\sim (\vdash (x \rightarrow y)) \rightarrow \sim (\vdash (x \wedge z \rightarrow y))$, where at least one variable that freely occurs in x and y does not occur in z.

The additional rule of inference: if $w \wedge (\vdash x) \vdash (\vdash y)$ and $w \wedge (\vdash z) \vdash (\vdash x)$, then $w \wedge (\vdash z) \vdash (\vdash v)$, where v is formed of z through substitution of y in place of at least one occurrence of x in z.

VII. On Complete (Rigorous) Induction

Our approach to induction is as follows. The theory of induction (inductive logic), as well as other sections of logic, only determine the properties of certain linguistic expressions for certain types of situations. By no means does it represent a theory of discoveries, indeed the logic of discoveries in general is nonsense. In many cases the poor efficiency of induction rules results from the fact that the actual conditions in which they are applied do not coincide with those that should be fixed in their formulation. The theory of induction is but an aspect of logic and not a special localized section of it. In principle, in various sections of logic the rules regarded as those of induction can be formulated. The theory of complete (rigorous) induction presented here is considered by us as a part of the theory of proof specifying the properties of the predicate of provability in association with the quantifier of generality for sets of ordered terms.

In what follows we shall consider sets of terms. In our discussion the variables for terms will be in turn regarded as terms. There is a relation between variables and terms: if a is a term from the set of values of a variable b, then $a \rightharpoondown b$; if b is a variable, a is a term and $a \rightharpoondown b$, then a is a term from the set of values of b. One can form pairs, triplets, etc., in general groups of terms (including variables). We shall agree that (1) an individual terms represents a group of terms; (2) if $a^1, \ldots, a^n$ are terms, then $(a^1, \ldots, a^n)$ is a group of terms. A group of terms is a term. For groups of terms the following rules hold:

$$((a^1,\ldots,a^n) \rightharpoondown (b^1,\ldots,b^n)) \dashv\vdash (a^1 \rightharpoondown b^1) \wedge \ldots \wedge (a^n \rightharpoondown b^n).$$
$$(\forall(a^1,\ldots,a^n))x \dashv\vdash (\forall a^1)\ldots(\forall a^n)x.$$
$$(\exists(a^1,\ldots,a^n))x \dashv\vdash (\exists a^1)\ldots(\exists a^n)x.$$

A group of terms $(a^1, \ldots, a^n)$ occurs as free in x if and only if all $a^1, \ldots, a^n$ occur as free in x. We shall call by a substitution of $(a^1, \ldots, a^n)$ in place of $(b^1, \ldots, b^n)$ in x the substitution of, respectively, a^1 in place of $b^1, \ldots, a^n$ in place of b^n everywhere, where $b^1, \ldots, b^n$ occur as free in x. Below, when speaking of terms (and variables) we shall have in mind groups of terms (groups of variables) as well, with the above remark taken into account.

The rules for individual terms represent a simple case of complete induction:

^{I}(1) $\quad v \wedge (\vdash x) \vdash (\vdash (\forall a)x)$,
(2) $\quad v \wedge (\vdash (\forall a)x) \vdash (\vdash x)$,

where v is an assertion 'A term a is individual'.

Assume that M is a finite set of terms. Suppose that it is decomposed into a finite number of non-empty and pairwise non-intersecting subsets $\mathrm{M}^1, \ldots, \mathrm{M}^n (n \geqslant 2)$. Let a be a variable for terms of M; ia is a variable for terms of $\mathrm{M}^i (i = 1, \ldots, n)$; ix is a statement formed of x through replacement of all free occurrences of a by ia with no ia's occurring in x; W is a statement '$(1a \rightharpoondown a) \wedge \ldots \wedge (na \rightharpoondown a)$, and for any term b the formula $(b \rightharpoondown a) \rightarrow ((b \rightharpoondown 1a) \vee \ldots \vee (b \rightharpoondown na))$' holds.

The rules of finite complete induction:

^{II}(1) $\quad \mathrm{W} \wedge (\vdash (\forall 1a)1x) \wedge \ldots \wedge (\vdash (\forall na)nx) \vdash (\vdash (\forall a)x)$.
(2) $\quad \mathrm{W} \wedge (\vdash (\forall a)x) \vdash (\vdash (\forall 1a)1x) \wedge \ldots \wedge (\vdash (\forall na)nx)$.

Corollaries of II.1 and II.2:

(3) $\quad \mathrm{W} \wedge (\sim(\vdash (\forall 1a)1x) \vee \ldots \vee \sim(\vdash (\forall na)nx)) \vdash \sim(\vdash (\forall a)x)$.
(4) $\quad \mathrm{W} \wedge \sim(\vdash (\forall a)x) \vdash \sim(\vdash (\forall 1a)1x) \vee \ldots \vee \sim(\vdash (\forall na)nx)$.

In a special case of finite complete induction each of $\mathrm{M}^1, \ldots, \mathrm{M}^n$ contains only one element. A still more special case results when each of terms occurring in $M^1, \ldots, M^n$ is individual. From I and II the following rules can be obtained:

$$W \wedge V \wedge (\vdash 1x) \wedge \ldots \wedge (\vdash nx) \vdash (\vdash (\forall a)x),$$
$$W \wedge V \wedge (\vdash (\forall a)x) \vdash (\vdash 1x) \wedge \ldots \wedge (\vdash nx),$$

where V is an assertion '$1a, \ldots, na$ are individual terms'.

Assume that M is a set of terms decomposed into non-empty and pairwise non-intersecting subsets M^1, M^2, M^3,.... Let these subsets be ordered in the following way: (1) M^1 is the first; (2) M^{i+1} immediately follows M^i, where $i = 1, 2, 3, \ldots$ (i.e., M^{i+1} is superior in order to M^i and none of M^1, M^2, M^3 is included in order between them). Suppose, further, that there are rules according to which one can construct (write, pronounce) all the elements of M^1 and construct all the elements of M^{i+1} if all the elements of M^i are specified. Assume that for any $M^i (i = 1, 2, 3, \ldots)$ it is possible to construct all the elements of M forming M^{i+1}, in other words, let the series of subsets $M^1, M^2, M^3, \ldots$ (and therefore the set M) be potentially infinite.

We shall introduce notations: a is a variable for terms of M; x is a statement in which a occurs as free; ia is a variable for terms of M^i with $(ia \to a)$ and ia does not occur in x; ix is a statement formed of x through substitution of ia in place of all free occurrence of a; W is a statement fixing the relations of ia and a, where $i = 1, 2, 3, \ldots$.

The rules of unconditional infinite complete induction:

III(1) $W \wedge (\vdash (\forall 1a) 1x) \wedge (\vdash ((\forall ia(ix \to (\forall (i+1)a)(i+1)x)) \vdash$
$\vdash (\vdash (\forall a)x),$

(2) $W \wedge (\vdash (\forall a)x) \vdash (\vdash (\forall 1a) 1x) \wedge (\vdash ((\forall ia) ix \to$
$\to (\forall (i+1)a)(i+1)x)).$

Corollaries of III:

(3) $W \wedge (\sim (\vdash (\forall 1a) 1x) \vee \sim (\vdash ((\forall ia) ix \to (\forall (i+1)a)(i+1)x))) \vdash$
$\vdash \sim (\vdash \forall a)x).$

(4) $W \wedge \sim (\vdash \forall ax) \vdash \sim (\vdash (\forall 1a) 1x) \vee \sim (\vdash ((\forall ia) ix \to$
$\to (\forall (i+1)a)(i+1)x)))).$

The conventional principle of mathematical induction represents a special case of this kind of complete induction, when each of $M^1, M^2, M^3, \ldots$ contains only one individual term.

Assume that p is a variable for terms such that the composition of M and its division into M^1, M^2, M^3 depends on the value of p. Suppose that the set of values of p has the following properties: (1) the first element is known (specified); (2) for any value of p one can construct its other value directly following it in order. Assume that for any value of p we can construct all elements of M^1, and if all elements of M^i are given, it is possible to construct all elements of M^{i+1}. The relations between $M^1, M^2, M^3, \ldots$ are the same as above. Let W^* fix the above relations of p, ia, a.

The rules of conditional infinite complete induction:

IV(1) $W^* \wedge (\vdash (\forall(p, 1a))1x) \wedge (\vdash ((\forall(p, ia))ix \to$
$\to (\forall(p, (i+1)a))(i+1)x) \vdash (\vdash (\forall(p, a))x)$.

(2) $W^* \wedge (\vdash (\forall(p, a))x) \vdash (\vdash (\forall(p, 1a))1x) \wedge (\vdash ((\forall(p, ia))ix \to$
$\to (\forall(p, (i+1)a))(i+1)x)$.

Corollaries of IV:

(3) $W^* \wedge (\sim(\vdash (\forall(p, 1a))1x) \vee \sim(\vdash ((\forall(p, ia))ix \to$
$\to (\forall(p, (i+1)a))(i+1)x)) \vdash \sim(\vdash (\forall(p, a))x)$.

(4) $W^* \wedge \sim(\vdash (\forall(p, a))x) \vdash \sim(\vdash (\forall(p, 1a))1x) \vee$
$\vee \sim(\vdash ((\forall(p, ia))ix \to (\forall(p, (i+1)a))(i+1)x))$.

Assume that D is a statement such that

$$D \dashv\vdash (z \in M),$$

where z is a variable for subject-terms. The statement D determines which exactly terms are included in M. Assume that $b^1, \ldots, b^m$ are all terms included in M^{i+1}, and $B^1, \ldots, B^m$ are all statements containing, respectively, $b^1, \ldots, b^m$ such that R is true:

$$\vdash (B^1 \wedge \ldots \wedge B^m \to (\forall(i+1)a)(i+1)x) \wedge ((\forall(i+1)a)(i+1)x \to \to B^1 \wedge \ldots \wedge B^m).$$

Let S be a statement

$$((\forall ia)ix \to B^1 \wedge \ldots \wedge B^k),$$

where $1 < k < m$. Suppose that $D^{k+1}, \ldots, D^m$ are the statements formed of D through substituting, respectively, $b^{k+1}, \ldots, b^m$ in place of all free occurrences of z. Assume that T is a statement

$$((\vdash ((\forall ia)ix \to B^{k+1})) \vee (\vdash D^{k+1} \to B^{k+1}))) \wedge \ldots$$
$$\ldots \wedge ((\vdash ((\forall ia)ix \to B^m)) \vee (\vdash (D^m \to B^m))).$$

The following rules of infinite complete induction with restriction hold:

V(1) $W \wedge R \wedge (\vdash (\forall 1a)1x) \wedge S \wedge T \vdash (\vdash (\forall a)x)$.
(2) $W \wedge R \wedge (\vdash \forall 1a)1x) \wedge S \wedge (\vdash (\forall a)x) \vdash T$.
VI(1) $W^* \wedge R \wedge (\vdash (\forall(p, 1a))1x) \wedge S \wedge T \vdash (\vdash (\forall(p, a))x)$.
(2) $W^* \wedge R \wedge (\vdash (\forall(p, 1a))1x) \wedge S \wedge (\vdash (\forall (p,a)x) \vdash T$.

Corollaries of V:

(3) $W \wedge R \wedge (\vdash (\forall 1a)1x) \wedge S \wedge \sim T \vdash \sim (\vdash (\forall a)x)$.
(4) $W \wedge R \wedge (\vdash (\forall 1a)1x) \wedge S \wedge \sim (\vdash \forall a)x) \vdash \sim T$.

Corollaries of VI:

(3) $W^* \wedge R \wedge (\vdash (\forall(p, 1a))1x) \wedge S \wedge \sim T \vdash \sim (\vdash (\forall (p,a)x)$.
(4) $W^* \wedge R \wedge (\vdash (\forall(p, 1a)1x) \wedge S \wedge \sim (\vdash (\forall(p,a)x) \vdash \sim T$.

VIII. On Logical Consistency

It is generally accepted that if in a logical theory such expressions as $x \wedge \sim x$ are provable, this theory is logically inconsistent and therefore inapplicable. We believe that such understanding of inconsistency of logical theories is not always suitable and acceptable. It is unsuitable when in a logical theory a class of rules of logical entailment of the form $x \vdash y^*$ is defined, and such formulas as $\sim (x \vdash y)$ do not appear at all. It is unacceptable when in a logical theory one has to take into account the structure of complex terms. According to the rules of construction of complex terms, it is possible to construct such terms that will allow us to prove both positive and negative statements containing a given term. For instance, consider a term 'A round square'. The statements 'A round square is round' and 'A round square is not round' are both provable, since every round subject is round, and every square is not round. In general, by applying the rules for construction of terms, we can construct terms such that the objects denoted by them can be assigned a certain attribute and can be denied it. How can one use this kind of term? Since up to now logic only infrequently dealt with this question and did not study the structure of terminology in detail, these problems were ignored, and logic tailored its solution in one or another way to suit the general opinion. It is impossible to forbid the construction of such terms, as it is impossible to forbid the construction of statements of the form $x \wedge \sim x$. One can just establish the rules according to which those should be treated, namely, consider these terms as logically contradictory. But in order to classify a term as logically contradictory, we must be able to deduce a logical contradiction from statements containing this term. In particular, to say that the term 'A round square' is logically contradictory, one has to be able to prove a statement 'The round square is round and the round square is not round', i.e., to prove a logically contradictory statement. From the traditional point of view this sounds nonsensical. But it is not. And there is no logical contradiction whatsoever in the very fact of proof of logical inconsistency of a statement containing a logically contradictory term. Evidently, in this situation one has to reject the universal nature of the semantic interpretation of consistency and inconsistency of logical theories and limit its application just to certain cases. It is also obvious that the general theory of logical entailment should exclude paradoxes similar to those of material and strict implication. In a number of works, we have considered logical theories of both kinds, trying to remain as long as possible within the traditional boundaries. In particular, we limited the logic of classes which contained structurally complex terms, in order to avoid the possibility of proof of contradictory statements, and used the semantic interpretation for the proof of consistency of the general theory of logical entailment. This was done, in particular, in our monographs *Foundations of the Logical Theory of Scientific Knowledge* (D. Reidel Publishing Co., Dordrecht-Holland, 1973), *Komplexe Logik* (Berlin, Braunschweig, Basel, 1970), 'The Logic of Classes' (a paper in the *Theory of Logical Inference* (collection of papers in Russian, Moscow, 1973). Here we shall give a brief formulation of the non-traditional understanding of the consistency of logical theories in the explicit form.

A propositional formula will be considered contradictory in the following and only in the following cases: (1) $p \wedge \sim p$ is contradictory (in this expression p is a propositional variable); (2) if A is contradictory, then $A \wedge B$ is contradictory; (3) if A and B are contradictory, then $A \vee B$ is contradictory; (4) if $x \vdash y$ is a theorem of the strong theory of logical entailment, and if y is contradictory, then x is contradictory.

A theory of logical entailment (i.e., a theory defining a class of logical theorems such as $x \vdash y$) will be regarded logically consistent if and only if a formula $A \vdash B$ such that A and B occur in a disjunctive normal form (in the form of disjunction of conjunctions of elementary formulas), B is contradictory by virtue of (1)–(3) of the above definitions and A is not contradictory according to these, is not provable (does not represent a theorem) in it.

It is proved in the theory of strong entailment that for any propositional formula x there is a formula y in a disjunctive normal form such that $x \vdash y$ and $y \vdash x$ are theorems. And the assertion of the consistency of the theory of strong entailment is proved as a usual metatheorem without applying the semantic interpretation.

The question of consistency of the theory of quantifiers is reduced to that of consistency of the theory of logical entailment by means of elimination of quantifiers and substitution of propositional variables in place of predicate formulas. In this situation an important fact is that at every stage of development of a logical theory (the logic of quantifiers, modalities, classes, etc.), the problem of the consistency of the corresponding sections should be solved by taking into account the peculiar features of this section. This is true in particular with respect to those chapters of logic that contain such theorems as $\vdash x$ (reads as 'x is accepted from purely logical considerations'). The theory of formulas of the form $\vdash x$ (the theory of degenerate entailment) is consistent at the level of propositional logic if and only if the formula A, where A is contradictory, is not a theorem in it. But at the level of logical theories dealing with the structure of terms, a different approach is required: the question of consistency of these theories should be reduced to that of the consistency of the theory of degenerate entailment. For example, if in the theory of terms the axioms

$$\vdash \mathrm{P}(a\downarrow \mathrm{P}),$$
$$(\wedge\ \mathrm{PQ})(a) \dashv\vdash \mathrm{P}(a) \wedge \mathrm{Q}(a),$$
$$\sim \mathrm{P}(a) \dashv\vdash \tilde{\mathrm{P}}(a),$$

are accepted, the following theorem

$$\vdash \mathrm{P}(a\downarrow(\wedge\ \mathrm{P}\tilde{\mathrm{P}})) \wedge \sim \mathrm{P}(a\downarrow(\wedge\ \mathrm{P}\tilde{\mathrm{P}})),$$

is provable in it. But that does not amount to a proof of the inconsistency of the theory of terms. The presence of such theorems in a theory makes it possible to introduce definitions of the inconsistency of terms and consider the question of the existence of corresponding objects. In particular, the term $(a\downarrow(\wedge\ \mathrm{P}\tilde{\mathrm{P}}))$ is contradictory, since the above corollary is provable in the theory of terms. In the section concerned with the predicates of existence one can obtain a proof of the assertion about the non-existence of an object denoted by such term. To forbid those 'contradictory' corollaries in the theory of terms means to limit the possibilities of logic without rational reasons for it (prejudices cannot be considered as grounds).

The corollaries lead to no misunderstanding of this kind since in our theory, in the general case, such formulas as

$$x \wedge \sim x \vdash y$$

are not theorems, i.e., any statement does not follow from a contradiction. Contradictions similar to the above remain just local phenomena playing an important role in the science of logic as a whole.

Certainly, one can introduce some limitations excluding such contradictory consequences (as was done by us, for example, in the logic of classes). For instance, instead of the axioms $\vdash \mathrm{P}(a\downarrow \mathrm{P})$ we can introduce the following ones: $\vdash \mathrm{P}(a) \rightarrow \mathrm{P}(a\downarrow \mathrm{P})$. And then all the contradictions of the above type can be regarded as a result of the fact that the antecedent is false. In this situation, of course, the respectability of logic will be preserved. But it will be self-deception and an unnecessary limitation and complication of logic. In one or another way the contradictory terms remain as a fact, and corresponding rules should be established for them.

The process of reduction discussed above consists in transforming the expressions in the theory of terms into predicate formulas and, finally, into propositional variables, and then in determining whether or not the resulting formulas contradict the fundamental theories, the theories of logical and degenerate entailment. For instance, such expressions as $(\wedge\ \mathrm{PQ})(a)$ transform in the process of such reduction into expressions of the form $\mathrm{P}(a) \wedge \mathrm{Q}(a)$, with the formulas

$$(\wedge\ \mathrm{PQ})(a) \dashv\vdash \mathrm{P}(a) \wedge \mathrm{Q}(a)$$

being transformed into expressions of the form

$$\mathrm{P}(a) \wedge \mathrm{Q}(a) \dashv\vdash \mathrm{P}(a) \wedge \mathrm{Q}(a),$$

and, finally, into formulas

$$p \wedge q \dashv\vdash p \wedge q.$$

The center of gravity of logical research is transferred to another question, more important for logic, namely, that of the correspondence of a logical theory to certain intuition. For example, a statement 'An object having an attribute P has this attribute' is intuitively indisputable. And it is absurd to limit it by the condition 'If an object has an attribute P'. The existence of negative predicates is also a fact. And the unification of terms according to the principle

$$P(a) \wedge Q(a) \dashv\vdash (\wedge\ PQ)(a),$$

where $(\wedge\ PQ)(a)$ reads as 'a has P and Q', is also a fact of linguistic practice.

On the basis of these considerations we reject the criticism that claims that our terms and classes that take into account the structure of terms are contradictory as unjust. The limitation of the logic of classes suggested by us in the above-mentioned works 'The Logic of Classes' and 'Logische Sprachreglen' we interpret as conceding to tradition. If in the logic of classes and terms, certain formulas such as

$$\vdash (A \rightharpoonup B) \wedge \sim(A \rightharpoonup B),$$
$$\vdash (A \in B) \wedge \sim(A \in B),$$

are proved, this does not constitute evidence of the inconsistency of those systems. It may be that this is evidence of the inconsistency of terms occurring in A.

IX. On the Systems of the Logic of Classes

1. *System SK*^{1}

The alphabet:

(1) K – the class-forming operator.

(2) $\in$ – the two-place predicate of the inclusion of individuals in a class.

D1. The subject form of SK^1:

(1) an individual variable is a subject form of SK^1;

(2) if a is a subject form of SK^1, then $\sim a$ is a subject form of SK^1;

(3) an arbitrary expression is a subject form of SK^1 only by virtue of 1 and 2.

(4) The elementary formula of SK^1: $\in(a, Kb)$ is an elementary propositional formula of SK^1 if and only if a and b are subject forms of SK^1.

The elementary propositional formula of SK^1 is a propositional formula. If x is a propositional formula, and a is a subject form of SK^1, then $(\forall a)x$ and $(\exists a)x$ are propositional formulas.

Instead of such symbols as $\in(a, Kb)$ we shall use the symbols, equivalent to them (but more visually obvious and more widely applied), of the form $(a \in \mathrm{K}b)$.

The axiomatic schemes of SK^1:

(A1) $(\exists a)(a \in \mathrm{K}b) \vdash (\exists b)(b \in \mathrm{K}a)$.
(A2) $\sim (a \in \mathrm{K}b) \vdash (a \in \mathrm{K} \sim b)$.
(A3) $(a \in \mathrm{K} \sim b) \vdash \sim (a \in \mathrm{K}b)$.
(A4) $(a \in \mathrm{K}b) \wedge (\forall b)(b \in \mathrm{K}c) \vdash (a \in \mathrm{K}c)$.
(A5) $(a \in \mathrm{K}b) \wedge (a \in \mathrm{K}c) \vdash (\exists b)(b \in \mathrm{K}c)$.

2. *System* SK^2

The alphabet:

(1) The list of variables for classes.
(2) K – the class-forming (the operator of primary classes).
(3) $\in$ – the two-place predicate of the inclusion of individuals in a class.
(4) $\subset$ – the two-place predicate of the inclusion of classes in a class.
(5) $\cup$ – the operator of unification of classes.
(6) $\cap$ – the operator of intersection of classes.
(7) $^{-}$ – the operator of complementation of classes.

The value-set of the variables for classes is formed by the terms of classes.

D1. The subject form of a class:

(1) the variable for a class is a subject form of a class;
(2) if a is an individual variable, then $\mathrm{K}a$ is a subject form of a class;
(3) if a is a subject form of a class, then $\bar{a}$ is a subject form of a class;
(4) if a and b are subject forms of a class, then $(a \cup b)$ and $(a \cap b)$ are subject forms of a class;
(5) an expression represents a subject form of a class only by virtue of (1)–(4).

The rules for omission and recovering of brackets for the subject forms of a class with operators $\cup$ and $\cap$ are similar to the corresponding rules for the propositional formulas with operators $\vee$ and $\wedge$.

D2. Subject forms of a class and individual variables are subject forms of SK^1.

D3. The elementary formula of SK^2:

(1) If a is an individual variable, and b is a subject form of a class, then $\in(a, b)$ is an elementary fromula of SK^2;

(2) if a and b are subject forms of a class, then $\subset(a, b)$ is an elementary formula of SK^2;

(3) an expression represents an elementary formula of SK^2 only by virtue of (1) and (2).

D4. An elementary formula of SK^2 is a propositional formula.

Instead of such symbols as $\in(a, b)$ and $\subset(a, b)$ we shall use, respectively, the symbols $(a \in b)$ and $(a \subset b)$, which are equivalent to them.

The axiomatic schemes of SK^2:

(A1) $(\exists a)(a \in \mathrm{K}b) \vdash (\exists b)(b \in \mathrm{K}a)$.
(A2) $\sim(a \in b) \vdash (a \in b)$.
(A3) $(a \in b) \vdash \sim(a \in b)$.
(A4) $(\mathrm{K}a \subset b) \vdash (\forall a)(a \in b)$.
(A5) $(\forall a)(a \in b) \vdash (\mathrm{K}a \subset b)$.
(A6) $(\forall a)((a \in b) \rightarrow (a \in c)) \vdash (b \subset c)$.
(A7) $(b \subset c) \vdash (\forall a)((a \in b) \rightarrow (a \in c))$.
(A8) $(a \in b) \vee (a \in c) \vdash (a \in (b \cup c))$.
(A9) $(a \in (b \cup c)) \vdash (a \in b) \vee (a \in c)$.
(A10) $(a \in b) \wedge (a \in c) \vdash (a \in (b \cap c))$.
(A11) $(a \in (b \cap c)) \vdash (a \in b) \wedge (a \in c)$.
(A12) $(a \in \mathrm{K}b) \wedge (\forall b)(b \in c) \vdash (a \in c)$.
(A13) $(a \in \mathrm{K}b) \wedge (a \in c) \vdash (\exists b)(b \in c)$.

3. *System* SK^3

The system SK^3 of the logic of classes results from making the following changes in the system SK^2.

An additional Definition D5 is adopted.

D5. The subject form of terms:

(1) an individual variable is a subject form of terms;

(2) if a is a subject form of terms, then $\sim a$ is a subject form of terms;

(3) if a and b are subject forms of terms, then $(a \vee b)$ and $(a \wedge b)$ represent subject forms of terms;

(4) an arbitrary expression is a subject form of terms only by virtue of (1)–(3).

An addition to the definition of a propositional formula: if a is a subject form of terms, and x is a propositional formula, then $(\forall a)x$ and $(\exists a)x$ are propositional formulas.

Paragraph 2 of Definition D1 is changed in the following way: (2) if a is a subject form of terms, then Ka is a subject form of classes.

An addition to D2: subject forms of terms are subject forms of SK^3.

Paragraph 1 of Definition D3 is changed as follows: (1) if a is a subject form of terms, and b is a subject form of classes, then $\in(a, b)$ is an elementary formula of SK^3.

The symbols SK^2 are replaced everywhere by the following ones:

(A1) $(\exists a)(a \in \mathrm{K}b) \wedge \sim(a \in \mathrm{K}(c \wedge \sim c)) \vdash (\exists b)(b \in \mathrm{K}a)$.
(A3) $(a \in b) \wedge \sim(a \in \mathrm{K}(c \wedge \sim c)) \vdash \sim(a \in b)$.
(A13) $(a \in \mathrm{K}b) \wedge (a \in c) \wedge \sim(a \in \mathrm{K}(d \wedge \sim d)) \vdash (\exists b)(b \in c)$.

Additional axiomatic schemes:

(A14) $(\forall a)(a \in \mathrm{K}b) \vdash (\forall \sim b)(\sim b \in \mathrm{K} \sim a)$.
(A15) $\vdash (a \subset \bar{\bar{a}}) \wedge (\bar{\bar{a}} \subset a)$.
(A16) $\vdash ((\overline{a \cup b}) \subset (\bar{a} \cap \bar{b})) \wedge ((\bar{a} \cap \bar{b}) \subset (\overline{a \cup b}))$.
(A17) $\vdash ((\overline{a \cap b}) \subset (\bar{a} \cup \bar{b})) \wedge ((\bar{a} \cup \bar{b}) \subset (\overline{a \cap b}))$.
(A18) $\vdash (\mathrm{K}(a \vee b) \subset (\mathrm{K}a \cup \mathrm{K}b)) \wedge ((\mathrm{K}a \cup \mathrm{K}b) \subset (\mathrm{K}(a \vee b))$.
(A19) $\vdash (\mathrm{K}(a \wedge b) \subset (\mathrm{K}a \cap \mathrm{K}b)) \wedge ((\mathrm{K}a \cap \mathrm{K}b) \subset \mathrm{K}(a \wedge b))$.
(A20) $\vdash (\mathrm{K} \sim a \subset \overline{\mathrm{K}a}) \wedge (\overline{\mathrm{K}a} \subset \mathrm{K} \sim a)$.

The following axiomatic schemes A^t are also adopted (symbols $x[\alpha/\beta]$ mean that the formula is obtained from x through substituting β in place of zero or more occurrences of α in x):

(1) $x \vdash x[\sim \sim a/a]$.
(2) $x \vdash x[a/\sim \sim a]$.
(3) $x \vdash x[a \wedge b/b \wedge a]$.
(4) $x \vdash x[a \wedge a/a]$.
(5) $x \vdash x[a/a \wedge a]$.
(6) $x \vdash x[a \wedge (b \wedge c)/(a \wedge b) \wedge c]$.
(7) $x \vdash x[(a \wedge b) \wedge c/a \wedge (b \wedge c)]$.
(8) $x \vdash x[(a \vee b) \wedge c/(a \wedge c) \vee (b \wedge c)]$.
(9) $x \vdash x[(a \wedge c) \vee (b \wedge c)/(a \vee b) \wedge c]$.

(10) $x \vdash x[\sim(a \wedge b)/\sim a \vee \sim b]$.
(11) $x \vdash x[\sim a \vee \sim b/\sim(a \wedge b)]$.
(12) $x \vdash x[a \wedge (b \vee \sim b)/a]$.
(13) $x \vdash x[a/a \wedge (b \vee \sim b)]$.

4. *System* SK^4

It would be interesting to consider the system SK^4 which is formed of SK^1 by making the following additions.

D1 is supplemented by a paragraph: if *a* and *b* are subject forms, then $(a \vee b)$ and $(a \wedge b)$ are subject forms.

Addition to the axiomatic schemes:

(A6) $(a \in \mathrm{K}b) \wedge (a \in \mathrm{K}c) \dashv\vdash (a \in \mathrm{K}(b \wedge c))$,
(A7) $(a \in \mathrm{K}b) \vee (a \in \mathrm{K}c) \dashv\vdash (a \in \mathrm{K}(b \vee c))$,
(A8) $\vdash a \in \mathrm{K}(a \vee b)$.

where *a*, *b*, *c* do not contain identical subject forms.

GLOSSARY OF SYMBOLS

GTD	the general theory of deduction
GTT	the general theory of terms
BA	base arithmetic
EA	expanded arithmetic
EP	empirical point
EL	empirical line
ES	empirical surface
EB	empirical body
SK	logic of classes
K	the operator of class
E	the predicate of existence
M	the predicate of possibility
N	the predicate of necessity
C	the predicate of contingency
R	the sign of dimension
l	the sign of dimension
W	the predicate of believe
Π	the predicate of preference
O	the predicate of realisation
f	the sign of function
$\rightharpoonup$	the sign of dependence by value
$\rightleftharpoons$	the sign of equivalence by value
$\equiv$	the sign of identity of meaning
$=$ Df.	equivalence of definition
$\equiv$ Df.	identity of meaning by definition
$\vdash$	the predicate of entailment and logical truth
$\dashv\vdash$	deductive equivalence
$\wedge$	conjunction
$\vee$	weakened disjunction
:	strong disjunction
$\sim$	external negation
$\rightarrow$	the conditionality operator
$\leftrightarrow$	the operator of reversible conditionality
$\urcorner$	internal negation
?	the uncertainty operator
$\forall$	the universal quantifier
$\exists$	the existence quantifier
$\downarrow$	the operator of limitation
$\mid$	the sign of adjacency
$\parallel$	the sign of contact
{ }	the sign of interval
$\Leftarrow$	the sign of correspondence
$\Lleftarrow$	the sign of one-to-one correspondence
$\in$	the sign of inclusion of an element in a set
$\subset$	the sign of inclusion of a class in a class
$\subset\supset$	the sign of equivalence of classes
$\Rightarrow$	the predicate of variation
$+$	the number-forming operator
$>$	the predicate of superiority
$<$	the predicate of inferiority
$=$	the predicate of identity
$\sqcup$	the relation 'between'
$\triangleright$	the sign 'therefore'
$\cup$	the operator of unification of classes
$\cap$	the operator of intersection of classes
$-$	the operator of complement of classes

SUBJECT INDEX

abstract objects 57
  – series 95
actual, the 77
adjacent 87

base arithmetic 115
beam 193
belief 259

cause 214
change 155
classes 60
  unification of – 64
  intersection of – 64
  complement of – 64
  correspondence of – 104
  logic of – 274
classes of numbers
  cardinality of – 130
classical case 31
clusters 63
complete induction 266
conjunction 12
connections 108
  paradoxes of – 207
contact 88
contingent events 73
continuity of space and time 162
contradiction 24
correspondence 102
  one to one – 105

deduction 25
  general theory of – 28
definitions
  implicit – 45
  incomplete – 46
  operational – 48
determinism 222
disjunction
  strong – 12
  weakened – 12

empirical body 200
  – connections 204
  – geometry 197
  – individuals 137
  – line 198
  – objects 57
  – point 197
empirical series
  continuity of an – 89
  discontinuity of an – 89
empirical surface 199
entailment
  predicate of 27
epistemic expressions 251
equivalence
  deductive – 28
estimates
  logic of – 262
events 65
existence 65
existential, the 77
expanded arithmetic 118
explication 55

fatalism 75
finite series 96
forecasts 237
functional dependence 107
functions 105

identity 81
implicit definitions 30
inclusion of an element in a class 61
indeterminism 222
individuals 59
inference
  logical – 26

inferiority 81
infinite numbers 119
infinite series 96
influence 210
interval 92
invariance of space and time 164

length 92
logical consistency 270
logical coordinates
theory of – 41
logical physics 136
logical truth
predicate of – 28

meaning
identity of – 13
metastatements 6
metaterms 6
microphysics 224
minimal dimensions 171
modal operator 77
modal predicates 69
– of contingency 69
– of necessity 69
– of possibility 69
motion
paradox of – 168
relativity of – 186
movement 167
negation
external – 12
internal – 12
non-classical case 31
norms
logic of – 262
numbers 112

occurrence of terms and statements 6
ontology 239
operator
conditionality – 12
statement-producing – 16
uncertainty – 12
operator of revisible conditionality 12
order relations 83
ordered series 86

part 235
possibility
measurement of – 78
potential, the 77
predicates
definition of – 15
process 170
proof
theory of – 265
pseudodefinitions 47

quantifier
universal – 12
quantifier of existence 12
quantity 125
questions
logic of – 262

rate 181
relation 'between' 85
relations 80
rules of logic 1

series
beginning of a – 89
end of a – 89
sets 60
space 140
standard classes 129
statements 11
coordinates of – 18
definitions with – 13
meaning of – 13
realizable 24
truth values of – 16
states 65
structure 97
superiority 81

tautology 24
tendencies
predicates of – 205
terms 2
complex – 4
consistency of – 56
definitions of – 9
general theory of – 33

terms (continued)
  meaning of – 8
  simple – 4
time 140
  irreversibility of – 159
trajectory 235

universe 194

value
  equality in – 3
  subordination by – 3
variables 29, 49
  definitions with – 53
variation 138
  predication of – 165
wave-particle duality 232
whole 235

Zeno's paradoxes 185